# Integer Partitions

The theory of integer partitions is a subject of enduring interest. A major research area in its own right, it has found numerous applications, and celebrated results such as the Rogers-Ramanujan identities make it a topic filled with the true romance of mathematics.

The aim of this introductory textbook is to provide an accessible and wide-ranging introduction to partitions, without requiring anything more of the reader than some familiarity with polynomials and infinite series. Many exercises are included, together with some solutions and helpful hints.

The book has a short introduction followed by an initial chapter introducing Euler's famous theorem on partitions with odd parts and partitions with distinct parts. This is followed by chapters titled Ferrers Graphs, The Rogers-Ramanujan Identities, Generating Functions, Formulas for Partition Functions, Gaussian Polynomials, Durfee Squares, Euler Refined, Plane Partitions, Growing Ferrers Boards, and Musings.

GEORGE E. ANDREWS is Evan Pugh Professor of Mathematics at the Pennsylvania State University. He is the author of many books in mathematics, including *The Theory of Partitions* (Cambridge University Press). He is a member of the American Academy of Arts and Sciences. In 2003, he was elected to the National Academy of Sciences (USA).

KIMMO ERIKSSON is a professor of applied mathematics at Mälardalen University in Sweden. He is the author of several matematical textbooks and popular articles in Swedish, as well as the opera *Kurfursten* with music by Jonas Sjöstrand. He is a member of the Västmanland Academy (Sweden).

To Joy and Charlotte

# Integer Partitions

GEORGE E. ANDREWS
*The Pennsylvania State University*

KIMMO ERIKSSON
*Mälardalen University*

CAMBRIDGE UNIVERSITY PRESS
Cambridge, New York, Melbourne, Madrid, Cape Town, Singapore,
São Paulo, Delhi, Dubai, Tokyo, Mexico City

Cambridge University Press
The Edinburgh Building, Cambridge CB2 8RU, UK

Published in the United States of America by Cambridge University Press, New York

www.cambridge.org
Information on this title: www.cambridge.org/9780521600903

First published 2004

*A catalogue record for this publication is available from the British Library*

*Library of Congress Cataloguing in Publication data*
Andrews, George E., 1938–
Integer partitions / George E. Andrews, Kimmo Eriksson.
p.  cm.
Includes bibliographical references and index.
ISBN 0-521-84118-6 – ISBN 0-521-60090-1 (pbk.)
1. Partitions (Mathematics)  I. Eriksson, Kimmo, 1967–  II. Title.
QA165.A55  2004
512.7′3 – dc22      2003069732

ISBN 978-0-521-84118-4 Hardback
ISBN 978-0-521-60090-3 Paperback

# Contents

# Preface

This is a book about integer partitions. If you have never heard of this concept before, we guess you will nevertheless be quite familiar with what it means. For instance, in how many ways can 3 be partitioned into one or more positive integers? Well, we can leave 3 as one part; or we can take 2 as a part and the remaining 1 as another part; or we can have three parts of size 1. This extremely elementary piece of mathematics shows that the answer to the question is: "There are three integer partitions of 3."

All existing literature on partition theory is written for professionals in mathematics. Now when you know what integer partitions are you probably agree with us that one should be able to study them without advanced knowledge of mathematics. This book is intended to fill this gap in the literature.

The study of partitions has fascinated a number of great mathematicians: Euler, Legendre, Ramanujan, Hardy, Rademacher, Sylvester, Selberg and Dyson to name a few. They have all contributed to the development of an advanced theory of these simple mathematical objects. In this book we start from scratch and lead the readers step by step from the really easy stuff to unsolved research problems. Our choice of topics was motivated by our desire to get to the meat of the subject directly. We wanted to move quickly to one of the most magnificent and surprising results of the entire subject, the Rogers-Ramanujan identitities. After that we introduce enough about generating functions to enable us to touch on the beginnings of the many fascinating aspects of the subject.

The intended audience is fairly broad. Obviously this should be the ideal textbook for a course on partitions for undergraduates. We have tried to keep the book to a modest length so that its topics would fit within one semester. Also there are often people with mathematical interests who do not have an advanced mathematics education. We hope these people will find this book tailor-made for them. Finally we believe that anyone with basic mathematics knowledge will find this book a solid introduction to integer partitions.

In order to make the text both easier and more stimulating to read, many arguments have been omitted and left as exercises at the end of sections. To many of these exercises you can find solutions and hints at the back of the book. The exercises vary in difficulty. We have tried to inform you of what to expect from the exercises by giving our estimate of the difficulty according to the following scale: 1 means straightforward, 2 means that you need a bit of problem solving skills, and 3 means that you are in for quite a challenge.

The idea for this book came up when the authors met at a conference in Philadelphia in 2000. Since then, all work has been carried out via mail and e-mail between Sweden and the United States. We are grateful to a number of people for assistance. Art Benjamin and Carl Yerger read the entire book, caught many mistakes, and made helpful suggestions. Kathy Wyland did some typing at Penn State. Brandt Kronholm, James Sellers and Ae Ja Yee read the galley proofs. Cambridge provided careful editorial advice, and Jim Propp made helpful and extensive comments on Chapter 11.

George E. Andrews
Kimmo Eriksson

# Chapter 1

## Introduction

Mathematics as a human enterprise has evolved over a period of ten thousand years. Rock carvings suggest that the concepts of small counting numbers and addition were known to prehistoric cavemen. Later, the ancient Greeks invented such things as rational numbers, geometry, and the idea of mathematical proofs. Arab and Chinese mathematicians developed the handy positional system for writing numbers, as well as the foundation of algebra, counting with unknowns. From the Renaissance and onward, mathematics has evolved at an accelerating pace, including such immensely useful innovations as analytical geometry, differential calculus, logic, and set theory, until today's fruitful joint venture of mathematics and computers, each supporting the other.

We will delve into, or at least touch upon, many of these modern developments – but really, this book is about mathematical statements of a kind that would have made sense already to the cavemen! One could imagine a petroglyph or cave painting of the following kind:

```
0__              0__  0__
 ||               ||   ||
0__
 ||


0__              0__0__0__
 ||               || || ||
0__
 ||
0__
 ||
```

```
0__0__0__          0__ 0__
 || || ||           ||  ||
                   0__
                    ||


0__                0__0__0__0__
 ||                 || || || ||
0__
 ||
0__
 ||
0__
 ||


0__0__0__          0__ 0__0__
 || || ||           ||  || ||
0__                0__
 ||                 ||
```

The concepts involved here are just small counting numbers, equality of numbers, addition of numbers, and the distinction between odd and even numbers. What is shown in the table is that for at least up to four animals, they can be lined up in rows of odd lengths in as many ways as in rows of different lengths. Written on today's blackboard instead of prehistoric rock, the table would have a more efficient design:

| | |
|---|---|
| $1 + 1$ | 2 |
| $1 + 1 + 1$<br>3 | 3<br>$2 + 1$ |
| $1 + 1 + 1 + 1$<br>$3 + 1$ | 4<br>$3 + 1$ |

The fact that there will always be as many items in the left column as in the right one was first proved by Leonhard Euler in 1748. But it is quite possible that someone observed the phenomenon earlier for small numbers, since it takes no more advanced mathematics than humans have accessed since the Stone Age. Nowadays, objects such as $3 + 1$ or $5 + 5 + 3 + 2$ are called *integer partitions*.

Stating it differently, an integer partition is a way of splitting a number into integer parts. By definition, the partition stays the same however we order the parts, so we may choose the convention of listing the parts from the largest part down to the smallest.

Euler's surprising result can now be given a more precise formulation: *Every number has as many integer partitions into odd parts as into distinct parts.* The table continues for five and six:

| | |
|---|---|
| $1 + 1 + 1 + 1 + 1$ | $5$ |
| $3 + 1 + 1$ | $4 + 1$ |
| $5$ | $3 + 2$ |
| $1 + 1 + 1 + 1 + 1 + 1$ | $6$ |
| $3 + 1 + 1 + 1$ | $5 + 1$ |
| $3 + 3$ | $4 + 2$ |
| $5 + 1$ | $3 + 2 + 1$ |

---

**EXERCISE**

1. Continue the table from seven up to ten and check for yourself that Euler was correct! See if you can obtain some intuition for why the numbers of integer partitions of the two kinds are always equal. (Difficulty rating: 1)

---

Statements of the flavor "every number has as many integer partitions of this sort as of that sort" are called *partition identities*. The above partition identity of Euler was the first, but there are many, many more. It is an intriguing fact that there are so many different and unexpected partition identities. Here is another, very famous, example: *Every number has as many integer partitions into parts of size 1, 4, 6, 9, 11, 14, ... as into parts of difference at least two.*

The numbers $1, 4, 6, 9, 11, 14, \ldots$ are best described as having last digit 1, 4, 6, or 9. Another way to put it is that when these numbers are divided by 5, the remainder is 1 or 4. Counting with remainders is called *modular arithmetic* and will appear several times in this book. In fact, it is striking that partition identities, their proofs and consequences, involve such a wide range of both elementary and advanced mathematics, and even modern physics. We hope that you will find integer partitions so compellingly attractive that they will lure you to learn more about these related areas too.

The last identity above was found independently by Leonard James Rogers in 1894 and Srinivasa Ramanujan in 1913. The tale of this identity is rich and has some deeply human aspects, one of which is that Rogers was a relatively unknown mathematican for a long time until the amazing prodigy Ramanujan rediscovered his results twenty years later, thereby securing eternal fame (at least among mathematicians) also for Rogers. The field of integer partitions comes with an unusually large supply of life stories and anecdotes that are romantic or astonishing, or simply funny. They are best presented, and best appreciated, in conjunction with the mathematics itself. Welcome to the wonderful world of integer partitions!

# Chapter 2
## Euler and beyond

In this chapter, we will show how identities such as Euler's, and many more, can be proved by the bijective method. However, although the bijective method is elegant and easy to understand, it is not the method Euler himself used. Euler worked with an analytic tool called *generating functions*, which is very powerful but demands a bit more mathematical proficiency. We will return to Euler's method in Chapter 5.

---

**Highlights of this chapter**

- We introduce basic set theory: union, intersection, and cardinality of sets.
- We show how bijections (one-to-one pairings of two sets) can be used to prove identities.
- A bijective proof of Euler's identity is given (the number of partitions into distinct parts equals the number of partitions into odd parts) using merging of equal parts, the inverse of which is splitting of even parts.
- Euler's identity is generalized to other "Euler pairs," that is, sets $M$ and $N$ such that the number of partitions into distinct parts in $M$ equals the number of partitions into parts in $N$.

---

## 2.1 Set terminology

We will need some concepts from set theory. In particular, a *set* is a collection of distinct objects, usually called *elements*. We can describe a set by listing its elements within curly brackets. For example, $\{1, 2, 4, 5\}$ is a set of four elements, all of which are integers. It is important to remember that the order of the elements implied by the list is not part of the set; thus the lists $\{4, 5, 2, 1\}$ and $\{1, 2, 4, 5\}$ describe the same set.

If you discard some elements of a set and retain the rest, you obtain a *subset*. The symbol $\subset$ means "is a subset of." For instance, $\{2, 5\} \subset \{1, 2, 4, 5\}$.

The *intersection* of two sets $N$ and $N'$ is the set of those elements that lie in both sets, denoted by $N \cap N'$. Two sets are *disjoint* if they have no element in common, that is, if their intersection is empty. The *union* of two sets $N$ and $N'$ is the set $N \cup N'$ containing all elements found in any or both of these sets. Thus if $N = \{1, 4\}$ and $N' = \{2, 4, 5\}$, then their intersection is $N \cap N' = \{4\}$ and their union is $N \cup N' = \{1, 2, 4, 5\}$. Intersections and unions are conveniently illustrated by so-called *Venn diagrams*, such as:

The number of elements in a set $N$ is denoted by $|N|$ and is often called the *cardinality* (or just the size) of the set.

---

**EXERCISE**

2. In the above example, we had $|N| = 2$, $|N'| = 3$, $|N \cap N'| = 1$, and $|N \cup N'| = 4$. It is no coincidence that $2 + 3 = 1 + 4$; in fact, for any sets $N$ and $N'$, it is always true that $|N| + |N'| = |N \cap N'| + |N \cup N'|$. Why? Draw the conclusion that the size of the union of two sets equals the sum of their respective sizes if, and only if, the two sets are disjoint. (Difficulty rating: 1)

---

## 2.2 Bijective proofs of partition identities

In order to formulate partition identities precisely and concisely, some notation is needed. Let $p(n)$ denote the number of integer partitions of a given number $n$. The function $p(n)$ is called the *partition function*. For example, we have $p(4) = 5$, since there are five partitions of the number four:

$$4, \quad 3 + 1, \quad 2 + 2, \quad 2 + 1 + 1 \quad \text{and} \quad 1 + 1 + 1 + 1.$$

In partition identities, we are often interested in the number of partitions that satisfy some condition. We denote such a number by $p(n \mid [\text{condition}])$. For example, Euler's identity takes the form

$$p(n \mid \text{odd parts}) = p(n \mid \text{distinct parts}) \quad \text{for } n \geq 1. \tag{2.1}$$

Now let us reflect a moment on how such an identity can be proved. For every single value of $n$, we can verify the identity by listing the partitions of both kinds, counting them, and finding the numbers to be equal. But the identity is stated for an infinite range of values of $n$, so we cannot verify it case by case; instead we must find some general argument that holds for each and every positive value of $n$. A natural idea would be to find a general way of counting the partitions, yielding an explicit expression, the same for both sides of the identity. In other words, if we could show that $p(n \mid \text{odd parts})$ equals, say, $n^2 + 2$ (or some other expression), and if we likewise could show that $p(n \mid \text{distinct parts})$ equals the same number, then we would of course have proved that the identity holds. But can we find such an expression for these functions? From the partition tables in the previous chapter, including Exercise 1, we can compute the first few values:

| $n$ | 1 | 2 | 3 | 4 | 5 | 6 | 7 | 8 | 9 | 10 |
|---|---|---|---|---|---|---|---|---|---|---|
| $p(n \mid \text{odd parts})$ | 1 | 1 | 2 | 2 | 3 | 4 | 5 | 6 | 8 | 10 |

The tabulated values do not seem to suggest any simple function such as a polynomial in $n$. Consequently this approach fails to prove the identity. But we fail because we try to accomplish more than we actually need! If we want to verify that the number of objects of a type X is equal to the number of objects of a type Y, then we do not need to find the actual numbers – it is enough to pair them up and show that every object of type X is paired with a unique object of type Y and vice versa. The "cave paintings" of Chapter 1 constitute such a pairing between partitions of $n$ into odd parts and partitions of $n$ into distinct parts, for $n = 2, 3, 4$. Such a one-to-one pairing between two sets is called a *bijection*. Hence, in order to prove a partition identity, we just need to find a bijection between the partitions in question.

$$x_1 \longmapsto y_1$$
$$x_2 \longmapsto y_2$$
$$x_3 \longmapsto y_3$$

Figure 2.1: A typical bijection between two sets of three elements.

It is not obvious what a bijection between partitions could look like. An integer partition of $n$ is just a collection of parts summing up to $n$, so a bijection between partitions must be described in terms of operations on parts. A simple operation is splitting an even part into two equal halves. The inverse of this operation is merging two equal parts into one part twice as large. This gives an immediate bijective proof of a partition identity:

$$p(n \mid \text{even parts}) = p(n \mid \text{even number of each part}) \quad \text{for } n \geq 1. \quad (2.2)$$

Study how the bijection works for $n = 6$:

$$6 \longmapsto 3 + 3$$
$$4 + 2 \longmapsto 2 + 2 + 1 + 1 \qquad (2.3)$$
$$2 + 2 + 2 \longmapsto 1 + 1 + 1 + 1 + 1 + 1$$

---

**EXERCISE**

3. For odd $n$, there can be no partitions into even parts, nor into parts with an even number of each size. Why? For even $n \geq 2$, find an alternative bijective proof of the above identity by finding bijections for each of the two equalities

$$p(n \mid \text{even parts}) = p(n/2) = p(n \mid \text{even number of each part}). \quad (2.4)$$

(Difficulty rating: 2)

---

## 2.3 A bijection for Euler's identity

Returning to Euler's identity, what must a bijection look like? It must have the property that when we feed it a collection of odd parts, it delivers a collection of distinct parts with the same sum. Its inverse must do the converse.

*From odd to distinct parts:* If parts are distinct, there are no two copies of the same part. Hence, if the input to the bijection contains two copies of a part, then it must do something about it. As we have seen above, a natural thing to do is to merge the two parts into one part of double size. We can repeat this procedure until all parts are distinct – since the number of parts decreases at every operation, this must occur at the latest when only one part remains. For example,

$$3 + 3 + 3 + 1 + 1 + 1 + 1 \mapsto (3 + 3) + 3 + (1 + 1) + (1 + 1)$$
$$\mapsto 6 + 3 + 2 + 2$$
$$\mapsto 6 + 3 + (2 + 2)$$
$$\mapsto 6 + 3 + 4.$$

*Tracing our steps back to odd parts:* The inverse of merging two equal parts is the splitting of an even part into two equal halves. Repeating this procedure must eventually lead to a collection of odd parts – since the size of some parts decreases at every operation, this must occur at the latest when all parts equal one. For example,

$$6 + 3 + 4 \mapsto 6 + 3 + (2 + 2)$$
$$\mapsto 6 + 3 + 2 + 2$$
$$\mapsto (3 + 3) + 3 + (1 + 1) + (1 + 1)$$
$$\mapsto 3 + 3 + 3 + 1 + 1 + 1 + 1.$$

It might seem that there is an arbitrariness in the order in which we choose to split (or merge) the parts. However, it is clear that splitting one part does not interfere with the splitting of other parts, so the order in which parts are split does not affect the result. Neither does the order of merging, since merging is the inverse of splitting.

Above, we have described a procedure of repeated merging of pairs of equal parts that we can feed any partition into odd parts, and that will result in a partition into distinct parts. Inverting every step gives a procedure of repeated splitting of even parts that takes any partition into distinct parts and results in a partition into odd parts. Hence, this procedure is a bijection proving Euler's identity. For $n = 6$, the bijection works as follows:

$$
\begin{array}{ll}
5 + 1 & \longmapsto 5 + 1 \\
3 + 3 & \longmapsto 6 \\
3 + 1 + 1 + 1 & \longmapsto 3 + 2 + 1 \\
1 + 1 + 1 + 1 + 1 + 1 \longmapsto 4 + 2
\end{array}
\tag{2.5}
$$

A common feeling among combinatorial mathematicians is that a simple bijective proof of an identity conveys the deepest *understanding* of why it is true. Test your own understanding on a few exercises!

---

**EXERCISES**

4. Why does the same bijection also prove the following stronger statement for $n \geq 1$?

$$p(n \mid \text{even number of odd parts}) = \tag{2.6}$$
$$p(n \mid \text{distinct parts, number of odd parts is even}), \tag{2.7}$$

as well as the same statement with both "even" changed to "odd." (Difficulty rating: 2)

5. In the bijection, we are merging pairs of equal parts. Change "pairs" to "triples"! If we merge triples of equal parts until no such triples remain, how can we describe the resulting partitions? The inverse would be to split parts

that are divisible by three into three equal parts. When does this process stop? What identity is proved by this new bijection? (Difficulty rating: 1)

6. Generalize the idea of the previous exercise and show that for any integers $k \geq 2$ and $n \geq 1$,

$p(n \mid$ no part divisible by $k) = p(n \mid$ less than $k$ copies of each part). (2.8)

(Difficulty rating: 1)

---

## 2.4 Euler pairs

The merging/splitting technique for proving Euler's identity is versatile. We can let it operate on other sets of partitions, say $A$ and $B$, as long as the splitting process takes all partitions in $A$ to partitions in $B$ and the merging process takes all partitions in $B$ to $A$.

For instance, let $A$ be the set of partitions of $n$ into parts of size one. The number of partitions in $A$ is $p(n \mid$ parts in $\{1\}) = 1$, since the only partition of $n$ satisfying the condition is the sum $1 + 1 + \cdots + 1$ of $n$ ones. The merging process will merge pairs of ones into twos, then merge pairs of twos into fours, then merge pairs of fours into eights, and so on until all parts are distinct. Consequently, the corresponding set $B$ must be the set of partitions of $n$ into distinct parts in $\{1, 2, 4, 8, \dots\}$ (powers of two). Now we must check that the splitting process will take every partition in $B$ to $A$. Clearly any power of two (say $2^k$) is split into a pair of powers of two ($2^{k-1} + 2^{k-1}$). Since the only power of two that is odd is $2^0 = 1$, the process will go on until all remaining parts are ones.

Hence, we have a bijection that proves that for any $n \geq 1$,

$p(n \mid$ parts in $\{1\}) = p(n \mid$ parts are distinct powers of two). (2.9)

And since the left-hand expression has the value one, we have proved that every positive integer has a unique partition into distinct powers of two. This is called the *binary representation* of integers. For example,

$$
\begin{aligned}
1 &= & & 2^0 = (1)_2 \\
2 &= & 2^1 & = (10)_2 \\
3 &= & 2^1 + 2^0 & = (11)_2 \\
4 &= 2^2 & & = (100)_2 \\
5 &= 2^2 + & 2^0 & = (101)_2 \\
6 &= 2^2 + 2^1 & & = (110)_2 \\
7 &= 2^2 + 2^1 + 2^0 & & = (111)_2,
\end{aligned}
\qquad (2.10)
$$

where $(b_k b_{k-1} \ldots b_0)_2$ is the number written with binary digits (bits). This is the common mode for computers to store numbers in memory.

We have now used the merging/splitting process in two different cases: first as a bijection proving Euler's indentity,

$$p(n \mid \text{parts in } \{1, 3, 5, 7, \ldots\}) = p(n \mid \text{distinct parts in } \{1, 2, 3, 4, 5, \ldots\}),$$
(2.11)

and then as a bijection proving the uniqueness of the binary representation,

$$p(n \mid \text{parts in } \{1\}) = p(n \mid \text{distinct parts in } \{1, 2, 4, 8, \ldots\}). \quad (2.12)$$

What are the limits of the versatility of the merging/splitting process? In other words, precisely for which sets $N$ of part sizes do we obtain a bijection to distinct parts in some set $M$? Let us call such a pair of sets an *Euler pair*. We can easily obtain new Euler pairs from old ones by just choosing a positive integer $c$ and multiplying every part by $c$. For example, multiplication of the parts in identity (2.9) by three yields the new Euler pair identity

$$p(n \mid \text{parts in } \{3\}) = p(n \mid \text{distinct parts in } \{3, 6, 12, 24, \ldots\}). \quad (2.13)$$

Let us now inspect the workings of the merging/splitting process again. Start with a collection of parts, all sizes of which are in a set $N$. Pairs of equal parts are merged and remerged until all remaining parts are distinct. The merging steps can be traced backward in a unique way by splitting even parts, if we just know when to stop splitting. Of course, we want to stop splitting when we have returned to parts with sizes in $N$. But if $N$ contains both, say, 3 and 6, then we wouldn't know if we should stop splitting at size 6, or if the original parts actually were of size 3.

$$6 + 6 \longrightarrow 12$$
$$3 + 3 + 3 + 3 \longrightarrow 6 + 6 \nearrow$$

Figure 2.2: If $N$ contains both 3 and 6, then a part of size 12 can emerge both as the result of merging $3 + 3 + 3 + 3$ and $6 + 6$. Then the process is not a bijection.

This problem occurs if, and only if, there are two elements in $N$ such that the first one is a power of two times the other one. Therefore, the merging/splitting process proves the following general Euler pair theorem:

**Theorem 1**

$$p(n \mid \text{parts in } N) = p(n \mid \text{distinct parts in } M) \quad \text{for } n \geq 1, \quad (2.14)$$

*where N is any set of integers such that no element of N is a power of two times
an element of N, and M is the set containing all elements of N together with
all their multiples of powers of two.*

The idea of this theorem was originally found by I. Schur, but it first appears
in full generality in Andrews (1969b). Schur never published his work on this
topic, and it appears in P. Bachmann's *Niedere Zahlentheorie* attributed to
"J." Schur. The initial "J." is also used in Volume II of L. E. Dickson's *History
of the Theory of Numbers*. This confusion led Andrews to refer to "I. J. Schur" in
many publications. However, Schur always published under the name "I. Schur."

---

**EXERCISE**

7. Let $\lfloor x \rfloor$ denote the largest integer smaller than or equal to $x$. Use Theorem
   1 to prove that $\lfloor n/3 \rfloor + 1$ is the number of partitions of $n$ into distinct parts
   where each part is either a power of two or three times a power of two.
   (Difficulty rating: 2)

---

We have now found the limits of applicability of the merging/splitting process
for proving Euler pair identities. One might wonder if there are any more Euler
pairs that we cannot find with this method. Let's investigate!

A typical set not covered by the theorem is $N = \{1, 3, 6\}$, since six equals
three times a power of two. We can try to construct a corresponding set $M$ such
that $(N, M)$ is an Euler pair.

| $n$ | 1 | 2 | 3 | 4 | 5 | 6 |
|---|---|---|---|---|---|---|
| $p(n \mid \text{parts in } \{1, 3, 6\})$ | 1 | 1 | 2 | 2 | 2 | 4 |

From the table, we can construct $M$ part by part so that $p(n \mid \text{parts in } N) =
p(n \mid \text{distinct parts in } M)$. Start by setting $M := \emptyset$.

1. There shall be one partition of 1. We have none using distinct parts in
   $M = \emptyset$, so 1 must be added to $M$.
2. There shall be one partition of 2. We have none using distinct parts in
   $M = \{1\}$, so 2 must be added to $M$.
3. There shall be two partitions of 3. We have only one $(2 + 1)$ using distinct
   parts in $M = \{1, 2\}$, so 3 must be added to $M$.
4. There shall be two partitions of 4. We have only one $(3 + 1)$ using distinct
   parts in $M = \{1, 2, 3\}$, so 4 must be added to $M$.

5. There shall be two partitions of 5. We have two $(3 + 2$ and $4 + 1)$ using distinct parts in $M = \{1, 2, 3, 4\}$, so 5 must not be added to $M$.

6. There shall be four partitions of 6. We have only two $(3 + 2 + 1$ and $4 + 2)$ using distinct parts in $M = \{1, 2, 3, 4\}$. If we add 6 to $M$, we obtain one extra partition but we cannot get another one.

We failed in step six. Since $M$ was uniquely constructed in the previous steps, there can be no other more successful alternative. Therefore, there can be no Euler pair with $N = \{1, 3, 6\}$. In fact, there can be no other Euler pairs than those given by Theorem 1. You are invited to prove this yourself in a short sequence of exercises.

---

### EXERCISES

8. For a given set $N$, there can be at most one set $M$ such that $(N, M)$ is an Euler pair. Why? Think backward: "If there were two different such sets, $M$ and $M'$, then there would have to be some smallest integer $n$ that lies in one set but not in the other." What does this mean for $p(n \mid$ distinct parts in $M)$ compared to $p(n \mid$ distinct parts in $M')$? (Difficulty rating: 2)

9. For a given set $N$ where some element is a power of two times some other element, say $2^k a$ and $a$, there can exist no set $M$ such that $(N, M)$ is an Euler pair. Why? Let $2^k a$ be the smallest element of the above mentioned type, and show that $M$ can be uniquely constructed so that it works successfully for all $n < 2^k a$ but that the construction will fail for $n = 2^k a$, just as in the above example. (Difficulty rating: 2)

10. Show that Euler pairs can be characterized more succinctly as pairs $(N, M)$ such that $2M \subset M$ and $N = M - 2M$. (Difficulty rating: 3)

11. (Andrews, 1969a) Show that the number of partitions of $n$ into $k$th powers $(k > 1)$ in which no part appears more than $k - 1$ times is always equal to 1. (Difficulty rating: 3)

---

# Chapter 3

## Ferrers graphs

A graphical representation of partitions is useful not only for the hypothetical cavemen of Chapter 1. Many amazing facts about partitions are best explained graphically.

---

**Highlights of this chapter**

- Ferrers graphs and Ferrers boards are two similar ways of representing an integer partition graphically: the parts of the partition are shown as rows of dots or squares, respectively.
- From a partition, we obtain its conjugate partition by exchanging rows and columns in the Ferrers graph.
- How fast does the partition function $p(n)$ grow? We show that it is bounded above by the famous Fibonacci numbers.
- An example of a nice proof of a partition identity using Ferrers graphs is Bressoud's bijection for the identity

  $p(n \mid \text{super-distinct parts})$
  $= p(n \mid \text{distinct parts, each even part} > 2 \cdot [\# \text{ odd parts}])$.

- A real classic proof in this tradition is Franklin's proof of Euler's pentagonal numbers theorem, which states that

  $p(n \mid \text{even} \# \text{ distinct parts}) = p(n \mid \text{odd} \# \text{ distinct parts}) + e(n)$,

  where $e(n)$ is 0 unless $n$ is a pentagonal number, $j(3j \pm 1)/2$ for some integer $j$, in which case $e(n) = (-1)^j$.

---

## 3.1 Ferrers graphs and Ferrers boards

There are two common ways of drawing such graphs for a partition, say $4 + 4 + 2 + 1 + 1$:

For some uses, the representation with dots looks best; we will call it the *Ferrers graph* of $4 + 4 + 2 + 1 + 1$. On other occasions, the representation with squares comes natural; we will call it the *Ferrers board* (or *Young diagram*) of $4 + 4 + 2 + 1 + 1$.

---

**EXERCISE**

12. From our example above, a formal definition of Ferrers graphs can be inferred. What conditions must a collection of rows of equidistant dots satisfy to be a Ferrers graph? (Difficulty rating: 1)

---

Much of this chapter will deal with various transformations on Ferrers graphs. If such a transformation is invertible, then it is a bijection and can be used for proving some partition identity. As a very elementary example of a transformation, take the Ferrers graph above and remove the top row.

We see that if we remove the top row from a Ferrers graph, we are left with a new Ferrers graph. If $r$ was the length of the removed row, then all rows of the new Ferrers graph have length less than or equal to $r$. Conversely, for any such Ferrers graph, we can add a row of length $r$ on the top and obtain a Ferrers graph. Thus we have a bijection proving the partition identity

$$p(n \mid \text{greatest part is } r) = p(n - r \mid \text{all parts} \leq r).$$

A variation of this technique that may make sense would be to remove the first column instead of the top row. We obtain a new Ferrers graph where no column is longer than the removed column. But the length of the first column is equal to the number of rows, that is, the number of parts of the partition. Hence this column-removal transformation proves the identity

$$p(n \mid m \text{ parts}) = p(n - m \mid \text{at most } m \text{ parts}).$$

Now take a new look at the shape of a Ferrers board. The contour consists of a horizontal line on top, a vertical line on the left, and a ragged sloping *rim* from the southwest corner to the northeast corner. Obviously, the information about the partition lies in the rim.

There are obviously two kinds of corners of the rim: *outer corners* look like ⌋ and *inner corners* look like ⌐. The two end-points of the rim are counted as inner corners. The outer corners are the squares that can be removed so that what remains is still a Ferrers board. Similarly, the inner corners are the places where squares can be added so that a new Ferrers board is formed. In Chapter 11, we will study Ferrers boards growing in this way.

---

**EXERCISES**

13. Do the positions of the outer corners suffice to determine the partition uniquely? (Difficulty rating: 1)

14. Do the positions of the inner corners suffice to determine the partition uniquely? (Difficulty rating: 1)

15. Suppose we enlarge a partition by filling each inner corner with a dot. Show that the number of inner corners is increased. (Difficulty rating: 2)

---

## 3.2 Conjugate partitions

Suppose you draw a Ferrers graph on a transparency and fumble so that it ends up face-down on the projector. (An all-too-common situation for a mathematics speaker.) Then the picture will still show a Ferrers graph! The fumbling speaker has performed the transformation called *conjugation*:

What has happened? The rows of the first graph have become the columns of the second graph, and vice versa. Consequently this transformation returns a partition with the *greatest part* equal to the *number of parts* of the original partition. This gives a bijective proof of yet another partition identity:

$$p(n \mid m \text{ parts}) = p(n \mid \text{greatest part is } m). \tag{3.1}$$

---

**EXERCISES**

16. Find the conjugate partition to (a) $3 + 3 + 2 + 2 + 1 + 1$, (b) $7 + 1$, (c) $5 + 4 + 2 + 2 + 1$. (Difficulty rating: 1)

17. For $n = 7$ and $m = 3$, explicitly show how conjugation proves Eq. (3.1) by listing all the pairings of partitions. (Difficulty rating: 1)

18. Use conjugation to give a complete bijective proof for the identity

$$p(n \mid \leq m \text{ parts}) = p(n \mid \text{all parts} \leq m). \tag{3.2}$$

(Difficulty rating: 1)

19. Complete the list of pairings from Exercise 17 to a list that explicitly shows the bijection for Eq. (3.2) for $n = 7$ and $m = 3$. (Difficulty rating: 1)

20. Partition identities usually involve conditions on *sizes of parts* and on the *number of parts*. We have seen how these notions are connected by conjugation. Another common condition is that parts be *distinct*. Use conjugation to show that

$$p(n \mid \text{distinct parts})$$
$$= p(n \mid \text{parts of every size from 1 to the largest part}). \tag{3.3}$$

For example, for $n = 6$, the left-hand side counts the four partitions $6$, $5 + 1$, $4 + 2$ and $3 + 2 + 1$ while the right-hand side counts the four partitions $1 + 1 + 1 + 1 + 1 + 1$, $2 + 1 + 1 + 1 + 1$, $2 + 2 + 1 + 1$, and $3 + 2 + 1$. (Difficulty rating: 2)

---

A partition is *self-conjugate* if it is its own conjugate. (A mathematician might say that self-conjugate partitions are the fixed points of the conjugation transformation.) For example, the self-conjugate partitions of 12 are $6 + 2 + 1 + 1 + 1 + 1$, $5 + 3 + 2 + 1 + 1$, and $4 + 4 + 2 + 2$:

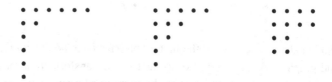

There is a natural transformation from self-conjugate partitions to partitions into distinct odd parts. Can you see it?

Take the first row together with the first column and make a new row of all those dots. Then take what is left of the second row and column and merge them into a new row, etc. Since self-conjugate partitions are symmetrical around the northwest-southeast diagonal, we always merge a row with a column of the same length – and since they share one dot, the result is a row twice the length of the original row less one, hence odd. It is obvious from the picture why the resulting parts must also be distinct.

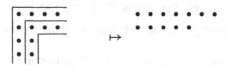

Conversely, starting from a partition into distinct odd parts, we can bend each odd part into a unique symmetric hook, and these hooks fit inside each other, forming a self-conjugate Ferrers graph. This bijection proves the identity

$$p(n \mid \text{self-conjugate}) = p(n \mid \text{distinct odd parts}). \qquad (3.4)$$

---

**EXERCISES**

21. Let us say that a partition is a *long rectangle* if its Ferrers graph is rectangular with length at least as great as height. Use the above idea of merging rows with corresponding columns to show that

    $$p(n \mid \text{consecutive parts differ by 2}) = p(n \mid \text{long rectangle}).$$

    (Difficulty rating: 1)

22. Show that $p(n \mid \text{long rectangle})$ equals the number of divisors of $n$ that are less than or equal to $\sqrt{n}$. (Difficulty rating: 1)

23. Show that the number of divisors of $n$ is odd if, and only if, $n$ is square. (Difficulty rating: 2)

24. Show that $p(n)$ is odd if, and only if, $p(n \mid$ distinct odd parts$)$ is odd. Use Eq. (3.4). (Difficulty rating: 2)

25. The *Durfee square* of a Ferrers board is the largest square that fits inside the board. For instance, the Durfee square of $4 + 4 + 2 + 1 + 1$ has size 2:

Observe that both below and to the right of the Durfee square sit smaller Ferrers boards. Try to prove that

$$p(n \mid \text{Durfee side} = j) = \sum_m p(m \mid \text{parts} \leq j) p(n - j^2 - m \mid \text{parts} \leq j).$$

We will return to Durfee squares in Chapter 8. (Difficulty rating: 2)

## 3.3 An upper bound on $p(n)$

How fast does the partition function $p(n)$ grow as $n$ grows? Like a polynomial, or an exponential function, or what? For that matter, how can we even be sure that $p(n)$ is a monotonically growing function, intuitive as it seems?

The complete answer to these questions is quite complex – in fact, we doubt that you have ever confronted such a complicated function as the partition function in your previous studies! We will take a glimpse at this rare monster in Chapter 5.

In the present section, we shall see that some playing around with Ferrers graphs gives a couple of interesting partial answers. Let us start out with proving that $p(n)$ really must grow as $n$ grows. In other words, we want to show that

$$p(n) > p(n - 1) \quad \text{for all } n \geq 2.$$

As an example, compare the partitions of three:

with the partitions of four:

$$\begin{matrix} \bullet & & \bullet\ \bullet & & \bullet\ \bullet\ \bullet & & \bullet\ \bullet & & \bullet\ \bullet\ \bullet\ \bullet \\ \bullet & & \bullet & & \bullet & & \bullet\ \bullet \\ \bullet & & \bullet & & & & \\ \bullet & & & & & & \end{matrix}$$

For every partition of $n - 1$, we obtain a partition of $n$ by adding a single dot in a new bottom row. Conversely, every partition of $n$ with a single dot in the bottom row gives a partition of $n - 1$ after we remove that dot. Hence, $p(n - 1) = p(n \mid$ at least one 1-part) and consequently

$$p(n) = p(n - 1) + p(n \mid \text{no 1-part}) > p(n - 1) \quad \text{for all } n \geq 2. \quad (3.5)$$

---

**EXERCISE**

26. List the Ferrers graphs for all partitions of five, and mark those that are left over in the pairing with partitions of four. (Difficulty rating: 1)

---

So $p(n)$ is a growing function. Our next task is to determine some bound on the growth speed, and our bound will be the famous *Fibonacci numbers*. These numbers were studied by the Italian mathematician Leonardo of Pisa (nicknamed "Fibonacci" – son of Bonaccio) in his book *Liber Abaci* (1202). Both Fibonacci and his numbers are exciting figures worthy of deeper study.

The Fibonacci numbers $F_0, F_1, F_2, \ldots$ are defined *recursively* by

$$F_0 = 0, F_1 = 1, \quad \text{and} \quad F_n = F_{n-1} + F_{n-2} \quad \text{for all } n \geq 2. \quad (3.6)$$

A definition is recursive when the same kind of object to be defined is used in the definition itself. In our case, the $n$th Fibonacci number is defined as the sum of the two previous Fibonacci numbers. Obviously, two initial numbers are needed to get this recursion going, so that is why the values of the first two Fibonacci numbers $F_0 = 0, F_1 = 1$ are part of the definition.

---

**EXERCISES**

27. Compute the first ten Fibonacci numbers. (Difficulty rating: 1)

28. Substituting $n - 1$ for $n$ in the definition yields $F_{n-1} = F_{n-2} + F_{n-3}$ for $n \geq 3$. Use this to show that (a)

$$F_0 = 0, F_1 = 1, F_2 = 1 \quad \text{and} \quad F_n = 2F_{n-1} - F_{n-3} \quad \text{for all } n \geq 3$$

and (b)

$$F_0 = 0, F_1 = 1, F_2 = 1 \quad \text{and} \quad F_n = 2F_{n-2} + F_{n-3} \quad \text{for all } n \geq 3$$

are valid alternative definitions of the Fibonacci numbers. (Difficulty rating: 2)

29. Show that $F_n = F_{n-1} + F_{n-3} + F_{n-5} + \ldots$ for $n$ even. (Difficulty rating: 2)

30. A *composition* of an integer is a partition where the order of the terms matters. For example, the compositions of four into ones and twos are $2 + 2, 2 + 1 + 1, 1 + 2 + 1, 1 + 1 + 2$, and $1 + 1 + 1 + 1$; five altogether. Note that also the fourth Fibonacci number is five! Show that for all $n \geq 1$, the number of compositions of $n$ into ones and twos equals $F_{n+1}$. (Difficulty rating: 2)

---

Now let us return to Eq. (3.5), which stated that $p(n) = p(n-1) + p(n \mid \text{no 1-part})$. In order to make a comparison with the recursion for Fibonacci numbers, we would like know how $p(n \mid \text{no 1-part})$ compares to $p(n-2)$.

First we observe that $p(n-2) = p(n \mid \text{at least one 2-part})$, since insertion/ removal of a 2-part is a bijection between the partitions in question. Second we note that we can transform any partition with no 1-part to a unique partition with at least one 2-part by cutting the smallest part (which is at least 2) into one 2-part and zero or more 1-parts.

For $n = 6$, the first bijection (insertion of a 2-part into a partition of $n - 2$) takes

to, respectively,

Merging all 1-parts and one 2-part into a new smallest part is possible for all but the second graph, yielding the four partitions of 6 with no 1-part:

The above argument actually shows that

$$p(n - 2) = p(n \mid \text{no 1-part})$$
$$+ p(n - 2 \mid \text{smallest non-1-part} < 2 + \# \text{1-parts}). \quad (3.7)$$

The last term is always nonnegative. Hence, combined with Eq. (3.5), this implies the Fibonacci-like inequality

$$p(n) \leq p(n-1) + p(n-2) \quad \text{for } n \geq 2. \tag{3.8}$$

Now we can prove the upper bound on $p(n)$ by mathematical induction.

**Theorem 2** *For all $n \geq 0$, the partition function $p(n)$ is less than or equal to the $(n+1)$st Fibonacci number $F_{n+1}$.*

*Proof.* Since $p(0) = F_1 = p(1) = F_2 = 1$, the proposition holds for $n = 0$ and $n = 1$. Assume that it holds for all $n < k$ for some $k \geq 2$. Then

$$\begin{aligned} p(k) &\leq p(k-1) + p(k-2) \quad &\text{(by Eq. (3.8))} \\ &\leq F_k + F_{k-1} \quad &\text{(by assumption)} \\ &= F_{k+1} \quad &\text{(by definition of } F_k\text{).} \end{aligned} \tag{3.9}$$

Hence, the proposition holds true for all $n \geq 0$.   Q.E.D.

---

**EXERCISES**

31. Compare the values of $p(n)$ and $F_{n+1}$ for $n = 1, 2, \ldots, 10$. Which is the first value of $n$ for which they differ? Why is that, in the light of the previous argument? (Difficulty rating: 1)

32. From the above proposition, we know that $p(n)$ grows at most as fast as the Fibonacci numbers – but how fast do *they* grow? Let $\tau$ denote the "golden mean" $(\sqrt{5}+1)/2$. Prove by induction that $\tau^{n-1} \leq F_{n+1} \leq \tau^n$ for all $n \geq 0$. (Difficulty rating: 2)

33. The most well-known combinatorial integer sequence after the Fibonacci numbers are the *Catalan numbers*, $C_m = \binom{2m}{m}/(m+1)$. By a *staircase* of height $m$, we mean the Ferrers board of the partition $1 + 2 + 3 + \cdots + (m-1)$, with empty steps added at the ends for convenience. For example, the staircase of height 4 looks as follows:

Verify that the number of Ferrers boards (including the empty board) fitting inside a staircase of height $m$ is $C_m$ for $n = 1, 2, 3, 4$. (Difficulty rating: 1)

34. The Catalan numbers obey the recursion

$$C_{m+1} = C_0 C_m + C_1 C_{m-1} + C_2 C_{m-2} + \cdots + C_m C_0 \quad \text{for } m \geq,$$

with $C_0 = 1$. Show that the number of Ferrers boards fitting inside a staircase of height $m$ obey the same recursion, by considering the lowest point $P$ of the staircase that is hit by the Ferrers board.

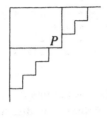

(Difficulty rating: 3)

---

## 3.4 Bressoud's beautiful bijection

Another way of describing that the numbers in a set are distinct is that every pairwise difference is at least one. Let us say that parts are *super-distinct* if every difference is at least two. We shall now present a relatively modern bijective proof of a partition identity concerning distinct and super-distinct parts.

Since every partition into super-distinct parts is also a partition into distinct parts, there must be more of the latter kind. For example, there are seven partitions of eleven into super-distinct parts: $11, 10 + 1, 9 + 2, 8 + 3, 7 + 4, 7 + 3 + 1, 6 + 4 + 1$. There are a few more partitions where the parts are distinct but not super-distinct: $8 + 2 + 1, 6 + 5, 6 + 3 + 2, 5 + 4 + 2$, and $5 + 3 + 2 + 1$. We shall prove the following strange-looking theorem:

**Theorem 3**

> $p(n \mid super\text{-}distinct\ parts)$
> $= p(n \mid distinct\ parts,\ each\ even\ part > 2 \cdot (\#\ odd\ parts))$.

For $n = 11$, the right-hand expression counts the seven partitions $11, 10 + 1, 8 + 3, 7 + 4, 7 + 3 + 1, 6 + 4 + 1$, and $6 + 5$. The bijective proof of David Bressoud (from 1980) goes as follows: Take the Ferrers graphs of partitions into super-distinct parts and adjust the left margin to a slope of two dots extra indentation per row. Draw a vertical line in such a way that the last row has one dot to the left of this line, the next-to-last line has three dots to the left, etc. Thus, for $14 + 11 + 6 + 4 + 1$, the graph looks like:

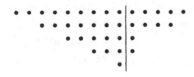

To the right of the line, we have obtained a new Ferrers graph, and we now rearrange its rows, starting with the odd rows in descending order followed by the even rows in descending order.

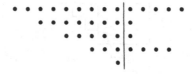

Ignoring the line and taking the rows of this graph as the parts of a partition, we get $14 + 8 + 6 + 7 + 1$, a partition into distinct parts with each even part greater than four (i.e., twice the number of odd parts).

---

**EXERCISES**

35. Explain why Bressoud's procedure always produces a partition into distinct parts with each even part greater than twice the number of odd parts. (Difficulty rating: 2)

36. Explain why the procedure is invertible, that is, why every partition into distinct parts with each even part greater than twice the number of odd parts is obtained exactly once. (Difficulty rating: 2)

37. Investigate what would have happened if the rows of the right-hand portion of the graph would instead have been reordered so that the even parts came first and the odd parts last (both in descending order). (Difficulty rating: 3)

38. Mini research project: Try to state and prove an analogous result for partitions into super-duper-distinct parts (where all differences are at least three)! (Difficulty rating: 3)

---

## 3.5 Euler's pentagonal number theorem

Euler was one of the most productive mathematicians in history. The total body of his collected work comprises many thousands of pages. Therefore, it should come as no surprise that there is more than one partition identity bearing his name; in addition to Euler's identity between the numbers of partitions into distinct parts and odd parts, we shall now present his identity between partitions into an odd resp. even number of distinct parts. Actually Legendre was the first to phrase Euler's result purely in terms of partitions. In contrast to all identities we have studied up to now, this one is not always perfectly accurate but has a correction term for certain numbers.

**Theorem 4 (Euler's pentagonal number theorem)**

$$p(n \mid even \# distinct\ parts) = p(n \mid odd \# distinct\ parts) + e(n) \quad (3.10)$$

*where $e(n) = (-1)^j$ if $n = j(3j \pm 1)/2$ for some integer $j$, and 0 otherwise.*

The name of the theorem is a good starting point for our discussion. The *triangular* numbers are $1, 3, 6, 10, \ldots$, referring to the number of dots in triangles of increasing size:

The $j$th triangular number is $1 + 2 + 3 + \cdots + j = j(j + 1)/2$. Similarly, the *pentagonal* numbers are $1, 5, 12, 22, \ldots$, referring to the number of dots in pentagons of increasing size:

Clearly, the $j$th pentagon consists of the $j$th triangle standing on top of a rectangle of width $j$ and height $j - 1$. Therefore, the $j$th pentagonal number is $j(j + 1)/2 + j(j - 1)$, which simplifies to $j(3j - 1)/2$. Now let us turn the pentagons on their side and adjust the dots in the triangle into straight rows, so that we obtain Ferrers graphs:

We see that these are Ferrers graphs of certain partitions into distinct parts: $1, 3 + 2, 5 + 4 + 3, 7 + 6 + 5 + 4$, etc. These particular partitions will appear as special cases in the following proof of Euler's pentagonal number theorem. This bijective proof was found by Franklin in 1881 and has achieved well-deserved fame.

We will try to create a bijection between partitions of some integer $n$ into an even number of distinct parts on the one side and partitions of $n$ into an odd number of distinct parts on the other side. Perfect for this purpose would be an invertible transformation that changes the number of parts by one, keeping the distinctness of parts. So, as a first idea, what happens if we take the smallest part and distribute its dots on the remaining rows, one on each row as far as they last? Watch three examples.

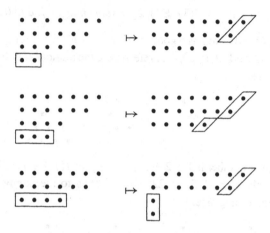

The transformation yields a partition into distinct parts if the rows were at least as many as the number of dots in the removed part – as in the first two examples. But we must demand something even stronger to make the transformation invertible, since the first two examples resulted in the same partition! Let's see what a sensible definition of the inverse could be.

In the inverse direction, we shall take a dot from a few of the largest parts and make a new smallest row. A well-defined number of dots to move would be the number of rows that differ by a single dot, starting with the largest row. In other words, we would remove the rightmost diagonal of the Ferrers graph.

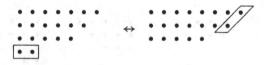

When should we remove the shortest row and when the rightmost diagonal? The only rule that makes sense is to move the rightmost diagonal if it is shorter than the shortest row, and otherwise move the latter.

However, there is a case when the above transformation fails to produce a valid Ferrers graph, namely when the shortest row actually intersects the rightmost diagonal in the lower right corner of the graph, and the row is the same length or one dot longer than the diagonal:

The Ferrers graphs in the first case are the pentagons of size $j(3j - 1)/2$ dots that we considered in the beginning of this section. The pentagons in the second case have an extra column in their rectangular part, giving a total size of $j(3j + 1)/2$ dots. We have described a transformation that *except for these pentagonal partitions*, pairs every partition of $n$ into an odd number of distinct parts with a partition of $n$ into an even number of distinct parts. Therefore,

$$p(n \mid \text{even \# distinct parts}) = p(n \mid \text{odd \# distinct parts}) + e(n),$$

where $e(n)$ is 0 unless $n = j(3j \pm 1)/2$ for some integer $j$, in which case it shall be 1 if the number $j$ of parts is even and $-1$ if odd. This proves Euler's pentagonal number theorem.

---

## EXERCISES

39. If $P$ is a partition of $n$, define the cardinality function $| \cdot |$ by $|P| = n$. Find a bijection that takes a pair $(P, P_{\text{even\#dist}})$ to a pair $(P', P_{\text{odd\#dist}})$, where $P$ and $P'$ are arbitrary partitions, $P_{\text{even\#dist}}$ and $P_{\text{odd\#dist}}$ are partitions into an even resp. odd number of distinct parts, and the total number of dots is preserved: $|P| + |P_{\text{even\#dist}}| = |P'| + |P_{\text{odd\#dist}}|$. (Difficulty rating: 3)

40. From the previous exercise, deduce that for $n \geq 1$ the sum

$$\sum_{k \geq 0} p(n - k)p(k \mid \text{even \# distinct parts})$$

equals the sum

$$\sum_{k \geq 0} p(n - k)p(k \mid \text{odd \# distinct parts}).$$

Use Euler's pentagonal theorem to draw the conclusion that

$$\sum_{j \geq 0} p(n - j(3j \pm 1)/2)(-1)^j = 0.$$

Explain how this formula can be interpreted as a recursion for the partition function:

$$p(n) = p(n - 1) + p(n - 2) - p(n - 5) - p(n - 7) + p(n - 12) + \ldots$$

This recursion will be alternatively derived and thoroughly discussed in Chapter 5. (Difficulty rating: 2)

41. Explain why Franklin's bijection also proves the following theorem of Fine: $p(n \mid \text{distinct parts, l.p. odd}) - p(n \mid \text{distinct parts, l.p. even}) = (-1)^j$ if $n = j(3j \pm 1)/2$, 0 otherwise. Here "l.p." stands for "largest part." (Difficulty rating: 2)

42. It is tempting to try Franklin's idea also on partitions into super-distinct parts, where the rightmost diagonal has half the slope. If you are not careful, you might come up with the incorrect conclusion that there are equally many such partitions with even and odd numbers of parts except for $n = 2j^2 - j, 2j^2, 2j^2 + j$. Explain both how one could be led to this erroneous conclusion and what oversight would have occurred! (Difficulty rating: 3)

43. Prove the partition identity

$$p(n \mid \text{even number of parts}) - p(n \mid \text{odd number of parts})$$
$$= (-1)^n p(n \mid \text{odd distinct parts})$$

by a bijection of Sylvester: Study the set of all partitions that are *not* into odd distinct parts, and pair them up by splitting the largest even part into two halves if these halves become larger than the previously largest repeated parts; otherwise, merge the two largest parts. Finish the argument by deciding the parity of the number of parts in a partition of $n$ into odd distinct parts. (Difficulty rating: 3)

44. Use the same bijection as in the previous exercise to prove that

$$p(n \mid \text{number of even parts is even}) - p(n \mid \text{number of even parts is odd})$$
$$= p(n \mid \text{odd distinct parts}).$$

Observe that the $(-1)^n$ factor from the previous exercise is no longer present! Why? (Difficulty rating: 2)

# Chapter 4
## The Rogers-Ramanujan identities

In this chapter, we will present some of the most famous and most spectacular partition identities ever found – the Rogers-Ramanujan identities and the seemingly related identity of Schur.

---

**Highlights of this chapter**

- Many partition identities, such as Euler's, are of the form that the number of partitions of some sort equals the number of partitions, where the parts belong to a certain set.
- Also, the two Rogers-Ramanujan identities are of this type, where the certain set consists of numbers congruent to $\pm 1$ (mod 5) and $\pm 2$ (mod 5), respectively. We will demonstrate how these identities can be "discovered."
- The form of the identities of Euler and Rogers-Ramanujan leads to Alder's conjecture, a still-open problem in the theory of integer partitions.
- Related to these results is the identity of Schur, where the certain set consists of numbers congruent to $\pm 1$ (mod 6). We present the clever bijective proof of Bressoud.

---

## 4.1 A fundamental type of partition identity

Many partition identities have the following fundamental structure: For some set $N$ of integers,

$$p(n \mid [\text{some condition}]) = p(n \mid \text{parts in } N) \quad \text{for all } n > 0. \quad (4.1)$$

We have already seen quite a few identities of this type with varying proofs. A few examples:

$$p(n \mid \text{even number of each part}) = p(n \mid \text{even parts}),$$

due to merging/splitting in Chapter 2. This is of type (4.1) with $N$ the set of even positive numbers. Chapter 2 ended with the more advanced identity

$$p(n \mid \text{distinct parts in } M) = p(n \mid \text{parts in } N) \quad \text{for Euler pairs } (N, M),$$

which is obviously of type (4.1). In Chapter 3, the conjugation transformation was used to prove

$$p(n \mid \le k \text{ parts}) = p(n \mid \text{parts} \le k).$$

This identity is also of type (4.1), with $N = \{1, 2, 3 \ldots, k\}$.

---

**EXERCISES**

45. Recall a partition identity that is *not* of type (4.1). (Difficulty rating: 1)
46. Prove the following type (4.1) identity:

$$p(n \mid \lambda_2 + \lambda_1, \text{ where } \lambda_2 \ge 2\lambda_1 \ge 0) = p(n \mid \text{parts in } \{1, 3\}).$$

Hint: Construct a bijection that results in $\lambda_1$ threes and a few ones. (Difficulty rating: 2)

47. As a minor variation on the previous exercise, prove

$$p(n \mid \lambda_2 + \lambda_1, \text{ where } \lambda_2 \ge 2\lambda_1 \ge 0, \lambda_2 \text{ is even}) = p(n \mid \text{parts in } \{2, 3\}).$$

(Difficulty rating: 2)

48. A slightly new twist! Prove

$$p(n \mid \lambda_2 + \lambda_1, \text{ where } \tfrac{3}{2}\lambda_1 \ge \lambda_2 \ge \lambda_1 \ge 0) = p(n \mid \text{parts in } \{2, 5\}).$$

(Difficulty rating: 2)

49. Find an analog to the previous identity for parts in $\{2, 3\}$. (Difficulty rating: 3)

## 4.2 Discovering the first Rogers-Ramanujan identity

The very first partition identity, Euler's identity, is clearly of type (4.1), with $N$ the set of odd positive numbers and the left-hand condition being that parts be distinct.

In exercises in Chapter 3, we generalized the distinctness condition to super-distinct and super-duper-distinct parts, meaning that every pair of parts differs by at least two and three, respectively. This nomenclature soon gets out of hand. For convenience, given any positive integer $d$, we coin the word *d-distinct* for parts that differ by at least $d$. For the special case $d = 0$, we define parts to be 0-distinct if there are at most two parts of any size.

We now construct a table over all partitions into 2-distinct parts of $n = 1, 2, \ldots, 11$.

| $n$ | # | partitions of $n$ into 2-distinct parts |
|---|---|---|
| 1 | 1 | 1 |
| 2 | 1 | 2 |
| 3 | 1 | 3 |
| 4 | 2 | $4, 3 + 1$ |
| 5 | 2 | $5, 4 + 1$ |
| 6 | 3 | $6, 5 + 1, 4 + 2$ |
| 7 | 3 | $7, 6 + 1, 5 + 2$ |
| 8 | 4 | $8, 7 + 1, 6 + 2, 5 + 3$ |
| 9 | 5 | $9, 8 + 1, 7 + 2, 6 + 3, 5 + 3 + 1$ |
| 10 | 6 | $10, 9 + 1, 8 + 2, 7 + 3, 6 + 4, 6 + 3 + 1$ |
| 11 | 7 | $11, 10 + 1, 9 + 2, 8 + 3, 7 + 4, 7 + 3 + 1, 6 + 4 + 1$ |

From the table we can, in a unique way, try to construct a set $N$ such that $p(n \mid$ parts in $N) = p(n \mid$ 2-distinct parts), much as we tried to construct Euler pairs in Chapter 2. Start by setting $N := \emptyset$.

1. There shall be one partition of 1. We have none using parts in $N = \emptyset$, so 1 must be added to $N$.

2. There shall be one partition of 2. We have one $(1 + 1)$ using parts in $N = \{1\}$, so 2 must *not* be added to $N$.

3. There shall be one partition of 3. We have one $(1 + 1 + 1)$ using parts in $N = \{1\}$, so 3 must *not* be added to $N$.

4. There shall be two partitions of 4. We have only one $(1 + 1 + 1 + 1)$ using parts in $N = \{1\}$, so 4 must be added to $N$.

5. There shall be two partitions of 5. We have two $(4 + 1$ and $1 + 1 + 1+$
   $1 + 1)$ using parts in $N = \{1, 4\}$, so 5 must *not* be added to $N$.
6. There shall be three partitions of 6. We have only two $(4 + 1 + 1$ and $1 +$
   $1 + 1 + 1 + 1 + 1)$ using parts in $N = \{1, 4\}$, so 6 must be added to $N$.
7. There shall be three partitions of 7. We have three $(6 + 1, 4 + 1 + 1 + 1,$
   and $1 + 1 + 1 + 1 + 1 + 1 + 1)$ using parts in $N = \{1, 4, 6\}$, so 7 must *not*
   be added to $N$.
8. There shall be four partitions of 8. We have four $(6 + 1 + 1, 4 + 4,$
   $4 + 1 + 1 + 1 + 1,$ and $1 + 1 + 1 + 1 + 1 + 1 + 1 + 1)$ using parts in
   $N = \{1, 4, 6\}$, so 8 must *not* be added to $N$.
9. There shall be five partitions of 9. We have four $(6 + 1 + 1 + 1, 4 + 4 + 1,$
   $4 + 1 + 1 + 1 + 1 + 1,$ and $1 + 1 + 1 + 1 + 1 + 1 + 1 + 1 + 1)$ using
   parts in $N = \{1, 4, 6\}$, so 9 must be added to $N$.

Thus the sequence of numbers that should be allowed parts starts $1, 4, 6, 9$.
Proceeding with the same argument, one finds that the sequence continues
$11, 14, 16, 19, 21, 24, \dots$. We always obtain two new numbers in the sequence
by adding five to the last two numbers computed. This set of numbers can
therefore be described as the set of positive integers that have remainder 1 or 4
when divided by 5. Such a number $m$ is said to be *congruent to* 1 *or* 4 *modulo*
5, with the mathematical notation

$$m \equiv 1 \text{ or } 4 \quad (\text{mod } 5).$$

Our investigation therefore puts us in a position to conjecture the following
partition identity:

$$p(n \mid \text{parts} \equiv 1 \text{ or } 4 \ (\text{mod } 5)) = p(n \mid 2\text{-distinct parts}). \qquad (4.2)$$

Identity (4.2) is the so-called *first Rogers-Ramanujan identity* that we mentioned
at the end of Chapter 1. We stress that our above argument does not prove this
identity for all $n$; we only verify it as far as we care to check. Unless a proof
valid for all $n$ is given, there is always the possibility that the identity fails to
hold for some still larger $n$.

Nevertheless, the above experimental method is a powerful tool for discov-
ering conjectural partition identities. Although this method was probably not
utilized by Ramanujan (who was much more involved with the generating function methods than with those directly involving partitions), it is almost certain
that Schur used it to discover the theorem we treat in Section 4.4. Although
doable by hand computation for small $n$, the method is very well suited for
computers. We invite you to try this method and rediscover a series of great
results in the theory of integer partitions.

**EXERCISES**

50. Rediscover the *second* Rogers-Ramanujan identity about partitions into 2-distinct parts greater than or equal to 2: Find a set $N$ such that $p(n \mid$ parts in $N) = p(n \mid$ 2-distinct parts $\geq 2)$. (Difficulty rating: 1)

51. Rediscover the first Göllnitz-Gordon identity about partitions into 2-distinct parts with no consecutive multiples of two (that is, if $2k$ is a part, then $2k + 2$ is not a part). (Difficulty rating: 2)

52. Rediscover the second Göllnitz-Gordon identity about partitions into 2-distinct parts $\geq 3$ with no consecutive multiples of two. (Difficulty rating: 2)

53. Rediscover Schur's identity about partitions into 3-distinct parts with no consecutive multiples of three. (Difficulty rating: 1)

54. Rediscover Gordon's identity about partitions into parts appearing at most twice, every part $\geq 2$; and if a part appears twice, then there is no part of an adjacent size. (Difficulty rating: 2)

55. Rediscover Andrews's identity about partitions into parts $\geq 2$, where every odd part is distinct and at least three greater than the next smaller part. (Difficulty rating: 2)

Methods utilizing computers to search for identities date back to the late 1960s. In a paper entitled *The use of computers in the search for identities of the Rogers-Ramanujan type* (Andrews, 1971b), an outline of such searches was first presented, culminating in a new version of Schur's theorem. Subsequently (Andrews, 1975), these search techniques were refined and more identities were discovered.

## 4.3 Alder's conjecture

The first Rogers-Ramanujan identity deals with parts that are congruent to 1 or 4 modulo 5. But 4 is congruent to $-1$ modulo 5, since $-1 + 5 = 4$. Thus we can more succinctly describe these partitions as having parts congruent to $\pm 1$ (mod 5). Now observe a fascinating pattern:

• By Exercise 5 in Chapter 2, partitions into parts not divisible by three are equinumerous with partitions into parts where every part appears at most twice. If an integer is not divisible by three, then division by three necessarily gives remainder 1 or 2 so that the integer is congruent to $\pm 1$ (mod 3). Hence, this identity can be expressed

$$p(n \mid \text{parts} \equiv \pm 1 \ (\text{mod } 3)) = p(n \mid \text{0-distinct parts}).$$

- By Euler's identity, partitions into odd parts are equinumerous with partitions into distinct parts. An integer is odd if, and only if, its remainder when divided by four is 1 or 3. Thus Euler's identity may be formulated

$$p(n \mid \text{parts} \equiv \pm 1 \ (\text{mod } 4)) = p(n \mid 1\text{-distinct parts}).$$

- The first Rogers-Ramanujan identity reads

$$p(n \mid \text{parts} \equiv \pm 1 \ (\text{mod } 5)) = p(n \mid 2\text{-distinct parts}).$$

The pattern above makes it very tempting to conjecture that every nonnegative integer $d$ gives an identity between the number of partitions into parts congruent to 1 or $-1$ (mod $d + 3$) and the number of partitions into $d$-distinct parts. The next case to test would be $d = 3$. Parts congruent to $\pm 1$ (mod 6) are 1, 5, 7, 11, etc. It is convenient to adopt a more concise notation, where the number of each part is registered in an exponent, so that $7 + 7 + 5 + 1 + 1 + 1 + 1$ is written $7^2 5^1 1^4$.

| $n$ | parts in $\{1, 5, 7, 11, \dots\}$ | 3-distinct parts |
|---|---|---|
| 1 | $1^1$ | 1 |
| 2 | $1^2$ | 2 |
| 3 | $1^3$ | 3 |
| 4 | $1^4$ | 4 |
| 5 | $1^5, 5^1$ | $5, 4 + 1$ |
| 6 | $1^6, 5^1 1^1$ | $6, 5 + 1$ |
| 7 | $1^7, 5^1 1^2, 7^1$ | $7, 6 + 1, 5 + 2$ |
| 8 | $1^8, 5^1 1^3, 7^1 1^1$ | $8, 7 + 1, 6 + 2$ |
| 9 | $1^9, 5^1 1^4, 7^1 1^2$ | $9, 8 + 1, 7 + 2, 6 + 3$ |

Did you notice what happened at $n = 9$? There were *three* partitions into parts congruent to $\pm 1$ (mod 6) but *four* partitions into 3-distinct parts! Thus our wonderful conjecture simply didn't hold up to careful scrutiny.

But perhaps we were just mistaken about the condition that parts be congruent to $\pm 1$ (mod $d + 3$)? There might possibly be some other set of parts that should be used, coinciding with parts congruent to $\pm 1$ (mod $d + 3$) for $d = 0, 1, 2$.

This turns out to be a vain hope. Lehmer proved that for any $d \geq 3$ there is no set $N$ such $p(n \mid d$-distinct parts) equals $p(n \mid \text{parts in } N)$ for all $n > 0$. Alder relaxed the original conjecture to an inequality:

$$p(n \mid \text{parts} \equiv \pm 1 \ (\text{mod } d + 3)) \leq p(n \mid d\text{-distinct parts}) \quad \text{for all } n, d \geq 0.$$

For values of $d$ that are a power of two less one (with the possible exception of 7), Andrews (1971a) has shown that the inequality always holds. Apart from this special case, Alder's conjectured inequality is still open to this day; neither proof nor disproof has been found. Large tables of these numbers seem to indicate that the inequality is sharp (that is, the "$\leq$" can be replaced with "$<$") for $n \geq d + 6 \geq 14$.

---

**EXERCISES**

56. Prove Lehmer's result for $d = 3$ and $d = 4$ by trying to construct the set $N$ and showing that such a try must fail. (Difficulty rating: 2)

57. What kind of transformation on partitions would you want to find in order to prove Alder's conjecture? (Difficulty rating: 2)

58. Write a computer program that computes the difference

$$p(n \mid d\text{-distinct parts}) - p(n \mid \text{parts} \equiv \pm 1 \ (\text{mod } d + 3))$$

and try to see some pattern in the data to come up with a refined conjecture of your own! (Difficulty rating: 3)

---

# 4.4 Schur's theorem

Separated by the First World War from the developments in British mathematics, Schur sat in Germany and made independent groundbreaking research on partitions. Not only did Schur find and prove the Rogers-Ramanujan identities, but in 1926 he also found the correct way of modifying the failed conjecture for $d$ equals three. The point is not to meddle with the parts congruent to $\pm 1$ (mod 6) but instead to exclude a certain subset of those partitions into 2-distinct parts.

**Theorem 5** *For any positive integer n, the number of partitions into parts $\equiv$ $\pm 1$ (mod 6) equals the number of partitions into 3-distinct parts where no consecutive multiples of 3 appear.*

This theorem brings new light on why the false conjecture

$$p(n \mid \text{parts} \equiv \pm 1 \ (\text{mod } 6)) = p(n \mid 3\text{-distinct parts})$$

did not fail until $n = 9$; this is the first number that can be partitioned into parts containing two consecutive multiples of 3, namely $6 + 3$.

In order to prove his theorem, Schur first recognized that the left-hand expression can be rewritten

$$p(n \mid \text{parts} \equiv \pm 1 \ (\text{mod } 6)) = p(n \mid \text{distinct parts} \equiv \pm 1 \ (\text{mod } 3)), \quad (4.3)$$

for the sets of positive integers $\{1, 5, 7, 11, \dots\}$ and $\{1, 2, 4, 5, 7, 8, 10, 11, \dots\}$ constitute an Euler pair. (Why? See Exercise 59.) Schur then gave a rather involved proof of the right-hand expression of this identity equals $p(n \mid 3\text{-distinct}$ parts, no consecutive multiples of 3). Several different proofs have been given since then by various mathematicians, and among them is the following compellingly clever bijection by Bressoud.

We begin with a partition $P$ into distinct parts congruent to $\pm 1$ (mod 3) and transform it into a new partition $P_1$ by merging pairs of parts differing by at most two, starting from the smallest part. It will be convenient to organize the parts into columns. For example,

$$P = \begin{matrix} 11 \\ 10 \\ 8 \\ 5 \\ 2 \\ 1 \end{matrix} \quad \mapsto \quad P_1 = \begin{matrix} 11 \\ 10+8 \\ 5 \\ 2+1 \end{matrix}.$$

The merged pairs always add to multiples of 3, and consecutive multiples of 3 cannot appear in this way. (Why?) Indeed, if there are $i$ numbers between two multiples of 3 in the column, then these multiples must differ by at least $(2+i)3$. (Why?)

Next we subtract consecutive multiples of 3 from the parts of $P_1$, starting with subtracting 0 from the bottom part and continuing upward. We leave the multiples of 3 in a new column at the side.

$$P_1 = \begin{matrix} 11 \\ 10+8 \\ 5 \\ 2+1 \end{matrix} \quad \mapsto \quad P_2 = \begin{matrix} 11-9 & 9 \\ 10+8-6 & 6 \\ 5-3 & 3 \\ 2+1-0 & 0 \end{matrix}.$$

We now rearrange the first column of $P_2$ in descending order.

$$P_2 = \begin{matrix} 11-9=2 & 9 \\ 10+8-6=12 & 6 \\ 5-3=2 & 3 \\ 2+1-0=3 & 0 \end{matrix} \quad \mapsto \quad P_3 = \begin{matrix} 12 & 9 \\ 3 & 6 \\ 2 & 3 \\ 2 & 0 \end{matrix}.$$

Finally we add up the numbers for each row in $P_3$.

$$P_3 = \begin{matrix} 12 & 9 \\ 3 & 6 \\ 2 & 3 \\ 2 & 0 \end{matrix} \quad \mapsto \quad P_4 = \begin{matrix} 21 \\ 9 \\ 5 \\ 2 \end{matrix} \; .$$

The last transformation must always result in a partition $P_4$ into 3-distinct parts where no consecutive multiples of 3 appear. (Why?) Thus we have described four simple transformations that combine to a transformation from a partition $P$ into distinct parts congruent to $\pm 1$ (mod 3) to a $P_4$ into 3-distinct parts with no consecutive multiples of 3. As we will discuss below, the combined transformation is also invertible. Hence it is a bijection, which proves Schur's theorem.

This bijection is very easily described in the direction treated above. However, giving a description of how the bijection works in the inverse direction is quite demanding. We leave the proof of invertibility as a series of exercises for the ambitious reader.

---

## EXERCISES

59. Show that $N = \{1, 5, 7, 11, \ldots\}$ and $M = \{1, 2, 4, 5, 7, 8, 10, 11, \ldots\}$ is an Euler pair. (Difficulty rating: 2)

60. Verify that the transformation from $P_1$ to $P_4$ is equivalent to the following game: As long as there exists some number that is not at least three greater than the number below, subtract three from this number, add three to the number below, and switch the positions of these numbers. Clearly, the game ends in a partition into decreasing 3-distinct parts. For example,

$$P_1 = \begin{matrix} \underline{11} \\ 18 \\ 5 \\ 3 \end{matrix} \quad \mapsto \quad \begin{matrix} 18+3=21 \\ 11-3=8 \\ \underline{5} \\ 3 \end{matrix} \quad \mapsto \quad \begin{matrix} 21 \\ \underline{8} \\ 3+3=6 \\ 5-3=2 \end{matrix}$$

$$\mapsto \quad \begin{matrix} 21 \\ 6+3=9 \\ 8-3=5 \\ 2 \end{matrix} = P_4$$

(Difficulty rating: 2)

61. Explain why the above game is inverted according to the following rules: First split parts of $P_4$ that are multiples of three into pairs of parts differing

by at most two.

$$P_4 = \begin{matrix} 21 \\ 9 \\ 5 \\ 2 \end{matrix} \quad \mapsto \quad P_4' = \begin{matrix} 11, 10 \\ 5, 4 \\ 5 \\ 2 \end{matrix}$$

This leaves a partition where no parts are multiples of three. Now, as long as the smallest part of some pair is less than three greater than the part below, subtract three from the largest part of the pair, add three to the part below, and switch their positions. Clearly, the game ends in a partition into parts that are not multiples of three, and where parts differing by at most two are paired up, starting from the smallest part.

$$P_4' = \begin{matrix} 11, 10 \\ \underline{5, 4} \\ 5 \\ 2 \end{matrix} \quad \mapsto \quad \begin{matrix} 11, 10 \\ 8 \\ 4, 2 \\ \underline{2} \end{matrix} \quad \mapsto \quad \begin{matrix} \underline{11, 10} \\ 8 \\ 5 \\ 2, 1 \end{matrix} \quad \mapsto \quad \begin{matrix} 11 \\ 10, 8 \\ 5 \\ 2, 1 \end{matrix} = P$$

(Difficulty rating: 3)

62. Combine your insights to explain why every partition into 3-distinct parts with no consecutive multiples of 3, and no others, can be obtained by the combined transformation starting from partitions into distinct parts congruent to $\pm 1 \pmod 3$. (Difficulty rating: 2)

---

Since there seem to be so many of these unexpected partition identities, one might as well expect the unexpected and look for more identities like Schur's theorem, say changing "3-distinct parts with no consecutive multiples of 3" to "2-distinct parts with no consecutive multiples of 2." And indeed, a theorem independently discovered by Göllnitz and Gordon says that the number of partitions with this condition equals the number of partitions into parts congruent to 1, 4, or 7 (mod 8).

But the obvious generalization does not hold. Alder proved that it is not possible, for any $d > 3$, to find a set $N$ such that

$$p(n \mid d\text{-distinct parts, no consecutive multiples of } d) = p(n \mid \text{parts in } N)$$

for all $n > 0$.

## EXERCISES

63. For $d = 1$, there is actually a set $N$ satisfying Alder's criterion. Which? (Difficulty rating: 2)

64. Prove Alder's result for $d = 4$ by trying to construct the set $N$ and showing that such a try must fail. (Difficulty rating: 1)

## 4.5 Looking for a bijective proof of the first Rogers-Ramanujan identity

We have now seen bijective proofs for a lot of identities. A bijection between two sets of partitions automatically yields a partition identity, but the reverse is far from true; given a partition identity, there is no routine procedure for constructing a bijection. The bijections presented earlier in this book have been of a variety of types, such as conjugation, splitting/merging, moving a diagonal in the Ferrers graph, organizing in two columns the internal orders of which are rearranged, etc.

To give a taste of the difficulties of finding a bijective proof for a partition identity known to be true, let us attack the first Rogers-Ramanujan identity. Below is a table of the sets that are equinumerous according to this identity, for $n = 1, 2, \ldots, 10$.

| $n$ | parts in $\{1, 4, 6, 9, 11, \ldots\}$ | 2-distinct parts |
|---|---|---|
| 1 | $1^1$ | 1 |
| 2 | $1^2$ | 2 |
| 3 | $1^3$ | 3 |
| 4 | $1^4, 4^1$ | $4, 3+1$ |
| 5 | $1^5, 4^1 1^1$ | $5, 4+1$ |
| 6 | $1^6, 4^1 1^2, 6^1$ | $6, 5+1, 4+2$ |
| 7 | $1^7, 4^1 1^3, 6^1 1^1$ | $7, 6+1, 5+2$ |
| 8 | $1^8, 4^1 1^4, 6^1 1^2, 4^2$ | $8, 7+1, 6+2, 5+3$ |
| 9 | $1^9, 4^1 1^5, 6^1 1^3, 4^2 1^1, 9^1$ | $9, 8+1, 7+2, 6+3, 5+3+1$ |
| 10 | $1^{10}, 4^1 1^6, 6^1 1^4, 4^2 1^2, 9^1 1^1, 6^1 4^1$ | $10, 9+1, 8+2, 7+3, 6+3+1, 6+4$ |

To produce a new row of this table, we can begin by taking every partition in the last row and adding a part of size 1 to partitions on the left, while adding 1 to the largest part of partitions to the right. Clearly, these new partitions satisfy the respective conditions. We then have to include partitions to the left that use only parts in $4, 6, 9, 11, \ldots$, and partitions to the right where the difference between the two largest parts is exactly two.

But this seems to be a promising beginning of a bijection! We now have an idea of how to take care of the parts of size 1 when transforming left-hand partitions – they shall all be merged and added to the largest part of what we obtain when transforming the rest of the parts into a partition into 2-distinct parts where the difference between the two largest parts is exactly two.

Next we must see how to transform the rest of the parts. From the table, we see that we must have

$$
\begin{aligned}
4 &\mapsto 3 + 1 \\
6 &\mapsto 4 + 2 \\
9 &\mapsto 5 + 3 + 1 \\
4^2 &\mapsto 5 + 3 \\
6^1 4^1 &\mapsto 6 + 4
\end{aligned}
$$

Clearly, parts of size 4 cannot all be transformed to $(3 + 1)$-blocks, for then $4^2$ would have been mapped to $6 + 2$. Instead it seems that the second 4 in $4^2$ and the 4 in $6^1 4^1$ have been mapped to a $(2 + 2)$-block. OK, we introduce a special rule for a single 4; all others are mapped as a $(2 + 2)$-block. We can now extend the table a few more rows and proceed in inferring what rules must govern the transformation of parts.

Countless mathematicians looking for a bijection for the first Rogers-Ramanujan identity have started out in this way. It looks promising at first, then more and more special rules must be added to the transformation procedure, and finally there seems to be nothing straightforward about it anymore. Time to give up.

The difficulty of describing a bijective explanation of the first Rogers-Ramanujan identity was recognized by the great peripatetic mathematician Paul Erdös. In addition to being the foremost mathematical problem solver of the twentieh century, Erdös loved to pose problems for others to attack. Depending on how difficult he judged the problem to be, Erdös set a price tag – ranging from $10 to $3,000 – for a published solution. Some of these problems were solved in Erdös's lifetime; he died in 1996 at age 83. For the remaining open problems, the offers are guaranteed by Ron Graham and Fan Chung, fellow mathematicians and longtime friends of Erdös, and Andrew Beal, banker and amateur mathematician.

For a bijective proof of the first Rogers-Ramanujan identity, Andrews (in imitation of Erdös) offered $50. Andrews has often wound up shelling out for the challenges he posed, and this was no exception. However, the published solution was not the kind of easily described bijection we have seen so many examples of in this book. The solution was published by Adriano Garsia and Stephen Milne in a paper fifty pages long, in which they invented a new method – the involution principle – for iteratively constructing bijections. David Bressoud and Doron Zeilberger used the same idea but managed to squeeze the proof down to two pages! Nevertheless, it is still an open and exciting question whether a direct bijection explaining the first Rogers-Ramanujan identity can be described.

As we have mentioned several times, the original discoverers of all these classic partition identities – Euler, Rogers, Ramanujan, Schur – used methods of proof other than bijections. But how is it possible to conclude that one set of partitions is as large as another set of partitions if you cannot exhibit a bijection between them? We shall see in the next chapter.

---

**EXERCISES**

65. Continue the above attempt at a Rogers-Ramanujan bijection by tabulating the partitions up to $n = 20$. (Difficulty rating: 1)

66. Find a complete bijective proof! (Difficulty rating: >3)

---

## 4.6 The impact of the Rogers-Ramanujan identities

While we have restricted ourselves to the intrinsic interest of the Rogers-Ramanujan and other partition identities, it would be a mistake to suggest they are of narrow interest. B. McCoy (in a joint paper with A. Berkovich (1998)) speaking to the International Congress of Mathematicians gave a survey of applications in physics. In addition, the fruitful interaction of partition identities with other combinatorial models is beautifully outlined by K. Alladi (1995), a work which presages further surprising and deep discoveries, K. Alladi et al. (2003).

# Chapter 5

## Generating functions

In the first four chapters, you were introduced to a purely arithmetic and combinatorial study of partitions. In Chapter 2, we presented an account of some of Leonhard Euler's discoveries from the eighteenth century. However, none of the methods used in that chapter were Euler's, only the results.

In fact, Euler's primary treatment of partitions was through the use of generating functions, the topic of this chapter.

---

### Highlights of this chapter

- Generating functions are power series designed to keep track of number sequences.
- Using generating functions, we can both state and prove Euler's theorem from Chapter 2 in another way.
- By restating Euler's pentagonal number theorem in terms of generating functions, we obtain a recursion for the partition function $p(n)$.
- The Rogers-Ramanujan identities can also be restated in terms of generating functions.

---

## 5.1 Generating functions as products

The idea here at its most elemental relies on the familiar rule for multiplying powers. Namely,

$$q^r \cdot q^s = q^{r+s}.$$

We may use this fundamental principle of algebra as follows. Suppose we wanted to exhibit all possible partitions consisting of one even part and one odd part, each $< 7$. We could write down each of them (there are nine); however, we note that they naturally arise in the following polynomial multiplication:

$$(q^2 + q^4 + q^6)(q^1 + q^3 + q^5)$$
$$= q^{2+1} + q^{2+3} + q^{2+5} + q^{4+1} + q^{4+3} + q^{4+5} + q^{6+1} + q^{6+3} + q^{6+5}$$
$$= q^3 + 2q^5 + 3q^7 + 2q^9 + q^{11}. \qquad (5.1)$$

Note that the second line clearly exhibits in the exponents all the partitions with one part even, one part odd, and each part $< 7$, and this fact is directly explained by (5.1) and the rules of multiplication of polynomials.

---

**EXERCISES**

67. Find the number of partitions of 9 into one odd and one even part by performing the polynomial multiplication

$$(q^1 + q^3 + q^5 + q^7)(q^2 + q^4 + q^6 + q^8)$$

and finding the coefficient of $q^9$. (Difficulty rating: 1)

68. Could you have seen the answer to Exercise 67 more directly? What is the number of such partitions of 10 and 11? (Difficulty rating: 1)

69. What polynomial $a_0 + a_1 q + a_2 q^2 + \cdots + a_r q^r$ has the property that $a_n = p(n \mid$ two parts; one $\leq 100$, one between 101 and 200) for every positive integer $n$. Describe it cleverly. (Difficulty rating: 1)

---

This simple idea may easily be expanded to much more general partition questions. For example, suppose $S = \{n_1, \ldots, n_r\}$ is a finite set of $r$ positive integers. When $r = 3$, we see that

$$(1 + q^{n_1})(1 + q^{n_2})(1 + q^{n_3})$$
$$= 1 + q^{n_1} + q^{n_2} + q^{n_3} + q^{n_1+n_2} + q^{n_1+n_3} + q^{n_2+n_3} + q^{n_1+n_2+n_3} \quad (5.2)$$

exhibits in the exponents all of the possible partitions using distinct elements of $\{n_1, n_2, n_3\}$.

To be even more explicit, if $S = \{1, 2, 3\}$, then the polynomial in (5.2) is

$$1 + q + q^2 + 2q^3 + q^4 + q^5 + q^6 \qquad (5.3)$$

This function (in this case a polynomial) is called the *generating function* for partitions into distinct elements of $\{1, 2, 3\}$; the coefficient of $q^n$ is the number of such partitions of $n$. Thus the coefficient of $q^3$ in (5.3) is 2, which corresponds

44          *Chapter 5. Generating functions*

to the fact that there are two partitions (3 and $2 + 1$) of 3 into distinct parts taken from $\{1, 2, 3\}$.

Exactly this reasoning allows us to conclude that if $S = \{n_1, n_2, \ldots, n_r\}$,

$$\sum_{n \geq 0} p(n \mid \text{distinct parts in } S)q^n = \prod_{i=1}^{r} (1 + q^{n_i}) = \prod_{n \in S} (1 + q^n). \quad (5.4)$$

Now suppose that we allow parts to repeat up to $d$ times. Say $d = 3$ and $S = \{n_1, n_2\}$

$$(1 + q^{n_1} + q^{n_1+n_1} + q^{n_1+n_1+n_1})(1 + q^{n_2} + q^{n_2+n_2} + q^{n_2+n_2+n_2})$$
$$= 1 + q^{n_2} + q^{n_2+n_2} + q^{n_2+n_2+n_2} + q^{n_1} + q^{n_1+n_2} + q^{n_1+n_2+n_2}$$
$$+ q^{n_1+n_2+n_2+n_2} + q^{n_1+n_1} + q^{n_1+n_1+n_2} + q^{n_1+n_1+n_2+n_2}$$
$$+ q^{n_1+n_1+n_2+n_2+n_2} + q^{n_1+n_1+n_1} + q^{n_1+n_1+n_1+n_2} + q^{n_1+n_1+n_1+n_2+n_2}$$
$$+ q^{n_1+n_1+n_1+n_2+n_2+n_2}$$
$$= \sum_{n \geq 0} p(n \mid \text{parts in } \{n_1, n_2\}, \text{no part repeated more than 3 times})q^n.$$

And in general, if $S = \{n_1, n_2, \ldots, n_r\}$, then

$$\sum_{n \geq 0} p(n \mid \text{parts in } S, \text{none repeated more than } d \text{ times})q^n$$

$$= \prod_{i=1}^{r} (1 + q^{n_i} + q^{n_i+n_i} + \cdots + q^{\overbrace{n_i + n_i + \cdots + n_i}^{d \text{ times}}})$$

$$= \prod_{i=1}^{r} (1 + q^{n_i} + q^{2n_i} + \cdots + q^{dn_i})$$

$$= \prod_{i=1}^{r} \frac{(1 - q^{(d+1)n_i})}{(1 - q^{n_i})} = \prod_{n \in S} \frac{1 - q^{(d+1)n}}{1 - q^n}, \quad (5.5)$$

where the final line in (5.5) has been obtained by application of the formula for the finite geometric series

$$\sum_{j=0}^{N} x^j = \frac{1 - x^{N+1}}{1 - x}. \quad (5.6)$$

Of course, we may wish to allow parts to appear arbitrarily many times (i.e., let $d \to \infty$). The argument in (5.5) is still valid, provided we require $|q| < 1$ (which is fine with us, since we use $q$ only for bookkeeping and not for its own value). Thus for $|q| < 1$,

$$\sum_{n \geq 0} p(n \mid \text{parts in } S)q^n = \prod_{i=1}^{r}(1 + q^{n_i} + q^{2n_i} + q^{3n_i} + \cdots)$$

$$= \prod_{i=1}^{r} \frac{1}{1 - q^{n_i}} = \prod_{n \in S} \frac{1}{1 - q^n}, \qquad (5.7)$$

where now we have invoked the sum of an infinite geometric series

$$\sum_{j=0}^{\infty} x^j = \frac{1}{1 - x}, \qquad |x| < 1.$$

The multiplication together of the $r$ infinite geometric series is completely legitimate because a finite number of absolutely convergent infinite series may be multiplied together and again yield an absolutely convergent infinite series.

---

**EXERCISE**

70. Why is

$$\frac{1}{(1 - q^5)(1 - q^{10})(1 - q^{25})}$$

the generating function for the number of ways of changing $n$ cents into nickels, dimes, and quarters? How does the generating function change if we also accept pennies and dollars? (Difficulty rating: 1)

---

Next we consider $S$ to be an infinite set of positive integers. Note that the extreme sides of each of (5.4), (5.5), and (5.7) make no reference to the finiteness of $S$. Thus we might hope that the same formulas hold when $S$ is infinite. In other words, for any set of $S$ of positive integers,

$$\sum_{n \geq 0} p(n \mid \text{distinct parts in } S)q^n = \prod_{n \in S}(1 + q^n),$$

$$\sum_{n \geq 0} p(n \mid \text{parts in } S, \text{no part repeated more than } d \text{ times})q^n$$

$$= \prod_{n \in S} \frac{\left(1 - q^{(d+1)n}\right)}{(1 - q^n)},$$

$$\sum_{n \geq 0} p(n \mid \text{parts in } S)q^n = \prod_{n \in S} \frac{1}{1 - q^n}, \qquad |q| < 1.$$

The proof that these identities are, in fact, correct will be reserved for the exercises below. Their plausibility is obvious in light of (5.4), (5.5), and (5.7). Appendix A can be consulted for reference to convergence properties of infinite products.

---

**EXERCISES**

71. In the following exercises, $S$ is an infinite set of positive integers arranged in increasing order:

$$S = \{n_1, n_2, n_3, n_4, \ldots\},$$
$$S_m = \{n_1, n_2, \ldots, n_m\}.$$

Prove that $p(n \mid \text{parts in } S_m) \leq p(n \mid \text{parts in } S)$. (Difficulty rating: 1)

72. Prove that for $n \geq n_m$, $p(n \mid \text{parts in } S_m) = p(n \mid \text{parts in } S)$. (Difficulty rating: 1)

73. Prove that

$$\lim_{m \to \infty} p(n \mid \text{parts in } S_m) = p(n \mid \text{parts in } S).$$

(Difficulty rating: 1)

74. Show that if

$$\mathcal{P}_m(q) = \sum_{n=0}^{\infty} p(n \mid \text{parts in } S_m)q^n = \prod_{i=1}^{m} \frac{1}{1 - q^{n_i}},$$

then

$$\mathcal{P}_m(q) = \sum_{n=0}^{n_m} p(n \mid \text{parts in } S)q^n + \sum_{n=n_m+1}^{\infty} p(n \mid \text{parts in } S_m)q^n$$

(Difficulty rating: 1)

75. Show that for $q$ real and $0 < q < 1$,

$$P(q) = \prod_{n \in S} \frac{1}{1 - q^n}$$

is an absolutely convergent infinite product. (See Appendix A, Fact 2.) (Difficulty rating: 2)

76. Show that

$$\sum_{n=0}^{n_m} p(n \mid \text{parts in } S)q^n < \mathcal{P}_m(q) < P(q).$$

(Difficulty rating: 2)

77. Prove that for $q$ real, $0 < q < 1$

$$\sum_{n=0}^{\infty} p(n \mid \text{parts in } S)q^n = \mathcal{P}(q).$$

(Difficulty rating: 2)

---

## 5.2 Euler's theorem

In Chapter 2, we proved a theorem of Euler asserting that

$$p(n \mid \text{parts all odd}) = p(n \mid \text{parts distinct}). \qquad (5.8)$$

Let us consider the related generating functions:

$$\sum_{n=0}^{\infty} p(n \mid \text{parts distinct})q^n = \prod_{n=1}^{\infty}(1 + q^n) \qquad \text{(by (5.4))},$$

and

$$\sum_{n=0}^{\infty} p(n \mid \text{parts all odd})q^n = \prod_{n \text{ odd}} \frac{1}{(1 - q^n)}.$$

Now clearly

$$\prod_{n=1}^{\infty}(1 + q^n)$$

$$= (1 + q)(1 + q^2)(1 + q^3)(1 + q^4)(1 + q^5)(1 + q^6) \cdots$$

$$= \left(\frac{1 - q^2}{1 - q}\right)\left(\frac{1 - q^4}{1 - q^2}\right)\left(\frac{1 - q^6}{1 - q^3}\right)\left(\frac{1 - q^8}{1 - q^4}\right)\left(\frac{1 - q^{10}}{1 - q^5}\right)\left(\frac{1 - q^{12}}{1 - q^6}\right) \cdots$$

$$= \frac{1}{(1 - q)(1 - q^3)(1 - q^5) \cdots} \qquad \begin{array}{l}\text{(by cancelling common factors} \\ \text{from the numerator and denominator)}\end{array}$$

$$= \prod_{n \text{ odd}} \frac{1}{1 - q^n}.$$

Thus the generating functions are identical; hence for every $n \geq 0$, Eq. (5.8) is true.

The algebraic manipulation of products can be used to prove countless theorems resembling (5.8).

For example, let us prove that

$p(n \mid$ each part $i$ appears $< i$ times$) = p(n \mid$ no parts are perfect squares$)$.

Let us look at an example. For $n = 9$, there are five partitions of the first type $(9, 7 + 2, 6 + 3, 5 + 4, 4 + 3 + 2)$ and five of the second type $(7 + 2, 6 + 3, 5 + 2 + 2, 3 + 3 + 3, 3 + 2 + 2 + 2)$.

The proof is patterned on the proof we just gave of (5.8).

$$\sum_{n \geq 0} p(n \mid \text{each part } i \text{ appears} < i \text{ times}) q^n$$

$$= \prod_{i=1}^{\infty} \left( 1 + q^i + q^{2i} + q^{3i} + \cdots + q^{(i-1)i} \right)$$

$$= \prod_{i=1}^{\infty} \frac{\left( 1 - q^{i^2} \right)}{\left( 1 - q^i \right)} \qquad \text{(by (5.6))}$$

$$= \prod_{\substack{n \\ \text{a non-square}}} \frac{1}{1 - q^n} = \sum_{n=0}^{\infty} p(n \mid \text{no parts are perfect squares}) q^n.$$

---

**EXERCISES**

78. (Subbarao, 1971b) Prove that the number of partitions of $n$ in which each part appears exactly 2, 3, or 5 times equals the number of partitions of $n$ into parts congruent to $\pm 2, \pm 3, 6 \pmod{12}$. (Difficulty rating: 2)

79. (MacMahon, 1916) Prove that the number of partitions of $n$ in which no part appears exactly once equals the number of partitions into parts not congruent to $\pm 1 \pmod 6$. (Difficulty rating: 3)

---

## 5.3 Two variable-generating functions

Sometimes we need to keep track of more than what number is being partitioned. In many instances, we want to have the generating function provide the number of parts of the partition as well.

Let us return to the example in Eq. (5.2); however, we shall now insert a second variable, $z$, whose exponent will keep count of how many parts are used in each partition:

$$(1 + zq^{n_1})(1 + zq^{n_2})(1 + zq^{n_3}) = 1 + zq^{n_1} + zq^{n_2} + zq^{n_3} + z^2 q^{n_1 + n_2}$$
$$+ z^2 q^{n_1 + n_3} + z^2 q^{n_2 + n_3} + z^3 q^{n_1 + n_2 + n_3}.$$

This example clearly leads us to the two variable analogs of (5.4), (5.5), and (5.7) following the exact same methods for only one variable:

$$\sum_{n \geq 0} \sum_{m \geq 0} p(n \mid m \text{ distinct parts each in } S) z^m q^n = \prod_{n \in S}(1 + zq^n), \quad (5.9)$$

$$\sum_{n \geq 0} \sum_{m \geq 0} p(n \mid m \text{ parts each in } S, \text{ each repeated} \leq d \text{ times}) z^m q^n$$

$$= \prod_{n \in S} \frac{\left(1 - z^{d+1} q^{(d+1)n}\right)}{1 - zq^n}, \quad (5.10)$$

$$\sum_{n \geq 0} \sum_{m \geq 0} p(n \mid m \text{ parts each in } S) z^m q^n = \prod_{n \in S} \frac{1}{1 - zq^n}. \quad (5.11)$$

If $S$ is infinite, then in each of (5.9), (5.10), and (5.11), we must require $|q| < 1$, and in (5.11) we must also require $|z| < \frac{1}{|q|}$.

## 5.4 Euler's pentagonal number theorem

In Chapter 3, we proved that

$$p(n \mid \text{even number of parts all distinct})$$
$$- p(n \mid \text{odd number of parts all distinct})$$
$$= \begin{cases} (-1)^j & \text{if } n = j(3j \pm 1)/2 \\ 0 & \text{otherwise.} \end{cases} \quad (5.12)$$

We may easily translate this assertion into one about the related generating function using Eq. (5.9). Namely, if we set $z = -1$ in (5.9), the resulting coefficient of $q^n$ on the left-hand side is precisely the left-hand side of (5.12). Therefore, by (5.9),

$$\sum_{n=0}^{\infty}(p(n \mid \text{even number of parts all distinct})$$
$$- p(n \mid \text{odd number of parts all distinct}))q^n$$

$$= \prod_{n=1}^{\infty}(1 - q^n) = \sum_{j=0}^{\infty}(-1)^j q^{j(3j-1)/2} + \sum_{j=1}^{\infty}(-1)^j q^{j(3j+1)/2}$$

$$= 1 + \sum_{j=1}^{\infty}(-1)^j q^{j(3j-1)/2}(1 + q^j) = \sum_{j=-\infty}^{\infty}(-1)^j q^{j(3j-1)/2}. \quad (5.13)$$

Now (5.13) is an extremely useful identity in that it provides us with an especially rapid algorithm for computing $p(n)$, the total number of all partitions of $n$. Namely, by (5.7),

$$\sum_{n=0}^{\infty} p(n)q^n = \prod_{m=1}^{\infty} \frac{1}{(1-q^m)}.$$

Therefore,

$$\prod_{m=1}^{\infty}(1-q^m)\sum_{n=0}^{\infty} p(n)q^n = 1;$$

so by (5.13),

$$\left(1+\sum_{j=1}^{\infty}(-1)^j q^{j(3j-1)/2}(1+q^j)\right)\sum_{n=0}^{\infty} p(n)q^n = 1. \qquad (5.14)$$

Comparing coefficients of $q^n$ on both sides of (5.14) for $n > 0$, we see that

$$p(n) - p(n-1) - p(n-2) + p(n-5) + p(n-7)$$
$$-\cdots+(-1)^j p\left(n-\frac{j(3j-1)}{2}\right)+(-1)^j p\left(n-\frac{j(3j+1)}{2}\right)$$
$$+\cdots = 0. \qquad (5.15)$$

If you are familiar with computer programming, you will see that (5.15) provides a very efficient algorithm for computing $p(n)$. In fact, $p(n)$ can be computed in a time proportional to $n^{\frac{3}{2}}$. Thus,

$$p(0) = 1$$
$$p(1) = p(0) = 1$$
$$p(2) = p(1) + p(0) = 2$$
$$p(3) = p(2) + p(1) = 3$$
$$p(4) = p(3) + p(2) = 5$$
$$p(5) = p(4) + p(3) - p(0) = 7$$
$$p(6) = p(5) + p(4) - p(1) = 11$$
$$p(7) = p(6) + p(5) - p(2) - p(0) = 15$$

etc.

## 5.5 Congruences for $p(n)$

When Hardy and Ramanujan were doing their research on $p(n)$ (discussed in the next chapter), they found that they needed a table of values of $p(n)$ in order to check their work. This was supplied by P. A. MacMahon, who made up a table of values for $p(n)$ with $1 \leq n \leq 200$. To make the table readable, he grouped the entries in blocks of five in the following manner:

| $n$ | $p(n)$ | $n$ | $p(n)$ | $n$ | $p(n)$ |
|---|---|---|---|---|---|
| 0 | 1 | 10 | 42 | 20 | 627 |
| 1 | 1 | 11 | 56 | 21 | 792 |
| 2 | 2 | 12 | 77 | 22 | 1002 |
| 3 | 3 | 13 | 101 | 23 | 1255 |
| 4 | 5 | 14 | 135 | 24 | 1575 |
| | | | | | |
| 5 | 7 | 15 | 176 | 25 | 1958 |
| 6 | 11 | 16 | 231 | 26 | 2436 |
| 7 | 15 | 17 | 297 | 27 | 3010 |
| 8 | 22 | 18 | 385 | 28 | 3718 |
| 9 | 30 | 19 | 490 | 29 | 4565 |

Ramanujan ("The Man Who Loved Numbers") noticed something that might pass right by the rest of us. Namely, the last $p(n)$ entry in each block is a multiple of 5. So he conjectured

$$p(5n + 4) \equiv 0 \quad (\text{mod } 5).$$

Given this completely unexpected possibility, he then tried other arithmetic progressions. Very soon he added the following conjectures:

$$p(7n + 5) \equiv 0 \quad (\text{mod } 7),$$
$$p(11n + 6) \equiv 0 \quad (\text{mod } 11).$$

Eventually he was able to prove each of these conjectures. More generally, he made comparable conjectures for any modulus of the form $5^\alpha 7^\beta 11^\gamma$. Numerous people worked on the problem, which was finally settled by G. N. Watson and A. O. L. Atkin; the latter's work was completed in 1969.

**EXERCISE**

80. Prove that

$$p(n \mid \text{all parts odd}) \equiv 0 \quad (\text{mod } 2)$$

except when $n = j(3j \pm 1)/2$. (Difficulty rating: 2)

## 5.6 Rogers-Ramanujan revisited

In Chapter 4, we introduced the Rogers-Ramanujan identities. From the work we have already done in this chapter, we see immediately that the generating function for partitions into parts congruent to $\pm 1$ (mod 5) is

$$\prod_{n=1}^{\infty} \frac{1}{(1 - q^{5n-4})(1 - q^{5n-1})},$$

whereas the generating function for partitions into parts congruent to $\pm 2$ (mod 5) is

$$\prod_{n=1}^{\infty} \frac{1}{(1 - q^{5n-3})(1 - q^{5n-2})}.$$

But what about partitions with super-distinct parts, i.e. partitions where the difference between parts is at least 2? Suppose $r_1 + r_2 + \cdots + r_m$ is such a partition, that is, $1 \leqq r_1 \leqq r_2 - 2$, $r_2 \leqq r_3 - 2$, etc. Then we may define $0 \leqq n_1 \leqq n_2 \leqq \cdots n_m$ uniquely by

$$r_1 = n_1 + 1$$
$$r_2 = n_2 + 3$$
$$r_3 = n_3 + 5$$
$$\vdots$$
$$r_m = n_m + 2m - 1.$$

Thus for any partition of $n$ with $m$ positive super-distinct parts, there is a corresponding partition of $n - (1 + 3 + \cdots + (2m - 1)) = n - m^2$ into $m$ non-negative parts (or equivalently at most $m$ positive parts). Hence, the generating function for partitions with $m$ positive super-distinct parts is

$$\sum_{0 \leq n_1 \leq n_2 \leq \cdots \leq n_m} q^{(1+n_1)+(3+n_2)+\cdots+(2m-1)+n_m}$$

$$= q^{m^2} \sum_{0 \leq n_1 \leq n_2 \leq \cdots \leq n_m} q^{n_1+n_2+\cdots+n_m}$$

$$= q^{m^2} \sum_{0 \leq n_1 \leq n_2 \leq \cdots \leq n_{m-1}} q^{n_1+\cdots+n_{m-1}} \frac{q^{n_{m-1}}}{1-q}$$

$$= \frac{q^{m^2}}{(1-q)} \sum_{0 \leq n_1 \leq n_2 \leq \cdots \leq n_{m-2}} q^{n_1+\cdots+n_{m-2}} \frac{q^{2n_{m-2}}}{1-q^2}$$

$$= \frac{q^{m^2}}{(1-q)(1-q^2)} \sum_{0 \leq n_1 \leq n_2 \leq \cdots \leq n_{m-3}} q^{n_1+\cdots+n_{m-3}} \frac{q^{3n_{m-3}}}{1-q^3}$$

$$\vdots$$

$$= \frac{q^{m^2}}{(1-q)(1-q^2)\cdots(1-q^m)}, \quad |q| < 1. \tag{5.16}$$

If we add up the expressions (5.16) for all $m \geq 0$, we will have the full generating function for all partitions with super-distinct parts, namely,

$$1 + \sum_{m=1}^{\infty} \frac{q^{m^2}}{(1-q)(1-q^2)\cdots(1-q^m)}.$$

Therefore, the Rogers-Ramanujan identities from Chapter 4 imply that

$$1 + \sum_{m=1}^{\infty} \frac{q^{m^2}}{(1-q)(1-q^2)\cdots(1-q^m)}$$

$$= \prod_{n=1}^{\infty} \frac{1}{(1-q^{5n-4})(1-q^{5n-1})}, \quad |q| < 1.$$

If we re-examine the above argument with the object of excluding 1 as a part, we see that we may now produce the $n_1, n_2, \ldots, n_m$ by

$$r_1 = n_1 + 2$$

$$r_2 = n_2 + 4$$

$$r_3 = n_3 + 6$$

$$\vdots$$

$$r_m = n_m + 2m$$

Thus in this case, the generating function is

$$1 + \sum_{m=1}^{\infty} \frac{q^{m^2+m}}{(1-q)(1-q^2)\cdots(1-q^m)}$$

because $m^2 + m = 2 + 4 + 6 + \cdots + 2m$.

Hence, the second Rogers-Ramanujan identity implies that

$$1 + \sum_{m=1}^{\infty} \frac{q^{m^2+m}}{(1-q)(1-q^2)\cdots(1-q^m)}$$

$$= \prod_{n=1}^{\infty} \frac{1}{(1-q^{5n-3})(1-q^{5n-2})}, \quad |q| < 1. \tag{5.17}$$

We will give a full proof of the Rogers-Ramanujan identities in Chapter 8.

# Chapter 6

## Formulas for partition functions

In Chapter 2, we noted that there are partition problems that could be solved easily if we had formulas for partition functions. In this chapter, we shall introduce an elementary method for finding formulas for some partition functions. We will really only probe the tip of this iceberg, and you will gain some idea of why this subject can become quite intricate.

---

**Highlights of this chapter**

- Study of generating functions gives formulas for $p(n, m)$, the number of partitions of $n$ into parts less than or equal to $m$, for $m = 1, 2, 3, 4$.
- As $n$ tends to infinity, $p(n)^{1/n}$ tends to 1.

---

## 6.1 Formulas for $p(n, 1)$ and $p(n, 2)$

Our primary focus will be on $p(n, m)$, the number of partitions of $n$ with each part $\leq m$. In terms of our previous notation,

$$p(n, m) = p(n \mid \text{parts in } \{1, 2, \ldots, m\}). \tag{6.1}$$

We see immediately that there is exactly one partition of a given $n$ into only 1s. So

$$p(n, 1) = 1.$$

It is not much more difficult to obtain a formula for $p(n, 2)$. Any partition of $n$ into 1s and 2s is uniquely determined by how many 2s are used, because once you know the number of 2s, the remaining parts are just the 1s. So for any $v$ with $0 \leq v \leq \frac{n}{2}$, there is a unique partition of $n$ into $v$ 2s and $n - 2v$, 1s. In other words,

$$p(n, 2) = \left\lfloor \frac{n}{2} \right\rfloor + 1,$$

where $\lfloor x \rfloor$ is the greatest integer $\leqq x$.

You can try the same idea for $p(n, 3)$, but things get somewhat tricky. The reasoning we just used on $p(n, 2)$ will quickly convince you that

$$p(n, 3) = \sum_{0 \leqq \nu \leqq \frac{n}{3}} p(n - 3\nu, 2) = \sum_{0 \leqq \nu \leqq \frac{n}{3}} \left( \left\lfloor \frac{n - 3\nu}{2} \right\rfloor + 1 \right). \qquad (6.2)$$

It turns out that there is a nice formula for $p(n, 3)$ (see next section), and if you are a glutton for punishment, you can prove this formula starting with the above sum. However, the ideas from Chapter 5 provide us with a powerful means for obtaining formulas for $p(n, m)$. This approach requires that you know the famous binomial series (a topic we return to in Chapter 7)

$$\sum_{n=0}^{\infty} \binom{n + m}{m} q^n = (1 - q)^{-m-1}, \quad |q| < 1, \qquad (6.3)$$

where

$$\binom{n + m}{m} = \frac{(n + 1)(n + 2) \cdots (n + m)}{m!} = \frac{(n + m)!}{n! \, m!}.$$

If $m = 0$, this is, of course, the summation of the geometric series

$$\sum_{n=0}^{\infty} q^n = \frac{1}{1 - q}, \quad |q| < 1. \qquad (6.4)$$

If we differentiate (6.4), we find

$$\sum_{n=0}^{\infty} n \, q^{n-1} = \frac{1}{(1 - q)^2}, \qquad (6.5)$$

which we may rewrite as

$$\sum_{n=0}^{\infty} \binom{n + 1}{1} q^n = (1 - q)^{-2},$$

which is (6.3) when $m = 1$.

If we differentiate (6.5) (i.e., if we take the second derivative of (6.4)), we find

$$\sum_{n=0}^{\infty} n(n - 1) q^{n-2} = \frac{2}{(1 - q)^3},$$

which we may rewrite as

$$\sum_{n=0}^{\infty} \binom{n+2}{2} q^n = (1-q)^{-3},$$

which is (6.3) when $m = 2$.

---

**EXERCISE**

81. Prove that (6.3) follows from (6.4) by differentiating $m$ times. (Difficulty rating: 2)

---

## 6.2 A formula for $p(n, 3)$

As we have seen in Chapter 5, $p(n, m)$ has the following generating function:

$$\sum_{n=0}^{\infty} p(n, m)q^n = \frac{1}{(1-q)(1-q^2)\cdots(1-q^m)}. \qquad (6.6)$$

What can we do with the product on the right-hand side of (6.6)? We want to alter it algebraically so that we can obtain tractable power series expansions. One first thinks of a partial fractions decomposition; for example,

$$\sum_{n=0}^{\infty} p(n, 3)q^n = \frac{1/6}{(1-q)^3} + \frac{1/4}{(1-q)^2} + \frac{17/72}{1-q} + \frac{1/8}{1+q} + \frac{1/9(q+2)}{1+q+q^2}.$$

It is possible but tedious and unpleasant to write Maclaurin series expansions for each of the five terms here. However, if we alter the partial fraction idea slightly, we see that

$$\sum_{n=0}^{\infty} p(n, 3)q^n = \frac{1/6}{(1-q)^3} + \frac{1/4}{(1-q)^2} + \frac{1/4}{1-q^2} + \frac{1/3}{1-q^3}$$

$$= \frac{1}{6} \sum_{n=0}^{\infty} \binom{n+2}{2} q^n + \frac{1}{4} \sum_{n=0}^{\infty} (n+1)q^n + \frac{1}{4} \sum_{n=0}^{\infty} q^{2n} + \frac{1}{3} \sum_{n=0}^{\infty} q^{3n}$$

$$= \sum_{n=0}^{\infty} \left( \frac{(n+3)^2}{12} q^n - \frac{1}{3}q^n + \frac{1}{4}q^{2n} + \frac{1}{3}q^{3n} \right)$$

$$= \sum_{n=0}^{\infty} \left( \frac{(n+3)^2}{12} + \epsilon(n) \right) q^n, \qquad (6.7)$$

where $\epsilon(n)$ takes only the values $-\frac{1}{3}, -\frac{1}{12}, 0, \frac{1}{4}$.

Now we can conclude by the uniqueness of Maclaurin series expansions that

$$p(n, 3) = \frac{(n+3)^2}{12} + \epsilon(n)$$

But $p(n, 3)$ is obviously an integer, and $|\epsilon(n)| < \frac{1}{2}$.

Consequently,

$$p(n, 3) = \left\{ \frac{(n+3)^2}{12} \right\}, \qquad (6.8)$$

where $\{x\}$ is the nearest integer to $x$.

This method dates back to Cayley and MacMahon, and has been extended by A. Munagi in his forthcoming Ph.D. thesis.

It is possible to derive (6.8) from (6.2), but it is very complicated. In addition, we see that as (6.7) evolved, we didn't know the formula (6.8) in advance. It just popped up at the end. This is one of the more powerful aspects of generating function techniques.

---

## EXERCISES

82. Show that $p(n, 2) = (2n + 3 + (-1)^n)/4$. (Difficulty rating: 2)

83. Show that

$$p(n \mid \text{parts in } \{1, 3, 5\}) = \left\{ \frac{(n+3)(n+6)}{30} \right\} = \left\{ \frac{(n+4)(n+5)}{30} \right\}$$

(Difficulty rating: 3)

84. Show that

$$p(n \mid \text{parts in } \{2, 3, 4\}) = \left\{ \frac{(n-3)^2}{12} \right\} - \left\lfloor \frac{n-3}{4} \right\rfloor \left\lfloor \frac{n-1}{4} \right\rfloor.$$

85. Show that the number of incongruent triangles with integer sides and perimeter $n$ is given by

$$p(n - 3 \mid \text{parts in } \{2, 3, 4\}).$$

(Difficulty rating: 3)

---

## 6.3 A formula for $p(n, 4)$

The simplicity of (6.8) is a little misleading as we look forward to considering $p(n, m)$ for $m > 3$. We treat $p(n, 4)$ in detail to reveal how the complications arise.

First of all, we shall require the following:

$$\sum_{n=0}^{\infty} a_n q^{2n} = \sum_{n=0}^{\infty} a_{n/2} \left( (n+1) - 2 \left\lfloor \frac{n+1}{2} \right\rfloor \right) q^n. \qquad (6.9)$$

The object of (6.9) is fairly simple. We want to replace the series on the left, which has only even exponents on $q$, with the series on the right, which has all nonnegative exponents. That we have accomplished this is immediate once we recognize that

$$(n+1) - 2 \left\lfloor \frac{n+1}{2} \right\rfloor = \begin{cases} 0 & \text{if } n \text{ is odd} \\ 1 & \text{if } n \text{ is even} \end{cases}$$

Following the example set by our previous treatment of $p(n, 3)$, we can start on $p(n, 4)$ with a partial fraction decomposition of the generating function:

$$\sum_{n=0}^{\infty} p(n, 4) q^n = \frac{1}{(1-q)(1-q^2)(1-q^3)(1-q^4)}$$

$$= \frac{1/24}{(1-q)^4} + \frac{1/8}{(1-q)^3} + \frac{59/288}{(1-q)^2} + \frac{17/72}{1-q} + \frac{1/32}{(1+q)^2}$$

$$+ \frac{1/8}{1+q} + \frac{(1+q)/9}{1+q+q^2} + \frac{1/8}{1+q^2}. \qquad (6.10)$$

Now (6.10) is quite unattractive and difficult to treat directly. However, a little algebra reveals that (6.10) may be altered to the following much more tractable formulation:

$$\sum_{n=0}^{\infty} p(n, 4) q^n = \frac{1}{(1-q)(1-q^2)(1-q^3)(1-q^4)}$$

$$= \frac{1/24}{(1-q)^4} + \frac{1/8}{(1-q)^3} + \frac{(5/12)^2}{(1-q)^2} + \frac{1/8}{(1-q^2)^2}$$

$$+ \frac{1/16}{1-q^2} + \frac{(2+q)/9}{1-q^3} + \frac{1/4}{1-q^4}$$

$$= \sum_{n \geq 0} \left( \left( \frac{1}{24} \binom{n+3}{3} \right) + \frac{1}{8} \binom{n+2}{2} + \left( \frac{5}{12} \right)^2 (n+1) \right) q^n$$

$$+ \left( \frac{1}{8}(n+1) + \frac{1}{16} \right) q^{2n}$$

$$+ \sum_{n \geq 0} \left( -\frac{1}{16} q^{2n} + \frac{2}{9} q^{3n} + \frac{1}{9} q^{3n+1} + \frac{1}{4} q^{4n} \right)$$

$$= \sum_{n \geq 0} \left( \frac{1}{24} \binom{n+3}{3} + \frac{1}{8} \binom{n+2}{2} + \left( \frac{5}{12} \right)^2 (n+1) \right) q^n$$

$$+ \left( \frac{1}{8} \left( \frac{n}{2} + 1 \right) + \frac{1}{16} \right) \left( (n+1) - 2 \left\lfloor \frac{n+1}{2} \right\rfloor \right) q^n$$

$$+ \sum_{n \geq 0} \left( -\frac{1}{16} q^{2n} + \frac{2}{9} q^{3n} + \frac{1}{9} q^{3n+1} + \frac{1}{4} q^{4n} \right) \qquad (6.11)$$

We now note that the power series represented by the final sum in (6.11) has coefficients that lie in the closed interval $[-\frac{1}{16}, \frac{17}{36}]$. That is, each of these coefficients is strictly less than $\frac{1}{2}$ in absolute value.

Consequently, given that $p(n, 4)$ is obviously an integer, we may conclude from (6.11) by the uniqueness of the Maclaurin series expansion that

$$p(n, 4) = \left\{ \frac{1}{24} \binom{n+3}{3} + \frac{1}{8} \binom{n+2}{2} + \frac{25}{144}(n+1) \right.$$

$$\left. + \frac{1}{8}(n+4) \left( \frac{n+1}{2} - \left\lfloor \frac{n+1}{2} \right\rfloor \right) \right\}$$

$$= \left\{ (n+1)(n^2 + 23n + 85)/144 - (n+4) \left\lfloor \frac{n+1}{2} \right\rfloor / 8 \right\}.$$

---

**EXERCISES**

86. Show that

$$p(n, 4) = \left\{ \left\lfloor \frac{n+4}{2} \right\rfloor^2 \left( 3 \left\lfloor \frac{n+9}{2} \right\rfloor - \left\lfloor \frac{n+10}{2} \right\rfloor \right) \right\} / 36 \right\}.$$

(Difficulty rating: 3)

87. Show that

$$p(n, 4) = \left\{ (n+5) \left( n^2 + n + 22 + 18 \left\lfloor \frac{n}{2} \right\rfloor \right) / 144 \right\}.$$

(Difficulty rating: 3)

88. Show that

$$p(n, 5) = \left\{ (n+8) \left( n^3 + 22n^2 + 44n + 248 + 180 \left\lfloor \frac{n}{2} \right\rfloor \right) / 2880 \right\}.$$

(Difficulty rating: 3)

---

Having clearly laid out for you the increasing complexities in treating $p(n, m)$, we conclude with some even more surprising descriptions of the deeper work on formulas for partition functions.

Perhaps the most famous result in the entire theory of partitions is the one found by G. H. Hardy, S. Ramanujan, and H. Rademacher for $p(n)$, namely,

$$p(n) = \frac{1}{\pi\sqrt{2}} \left( \frac{d}{dx} \frac{\sinh\left(\pi \left(\frac{2}{3}(x - 1/24)\right)^{\frac{1}{2}}\right)}{\left(x - \frac{1}{24}\right)^{\frac{1}{2}}} \right)_{x=n} + \text{similar terms.} \quad (6.15)$$

The proof of (6.15) and related formulas relies on an extremely subtle study of the generating function for $p(n)$ involving the power of the theory of functions of a complex variable.

The formulas of the previous sections, along with relevant history, are taken from Andrews (2003).

## 6.4 $\lim_{n \to \infty} p(n)^{1/n} = 1$

Although we cannot prove (6.15) or anything remotely like it in this text, we can say something about $p(n)$, namely, that it satisfies

$$\lim_{n \to \infty} p(n)^{1/n} = 1 \quad (6.16)$$

and so may be called a subexponential function.

Even the proof of (6.16) is somewhat intricate. First, we shall prove

$$np(n) = \sum_{j=1}^{n} p(n - j)\sigma(j), \quad (6.17)$$

where $\sigma(j)$ is the sum of the divisors of $j$. To see (6.17), we merely write down all the partitions of $n$ and then add them all up. Since there are $p(n)$ of them, the total of this sum must be $np(n)$. On the other hand, let us keep track of how many times the summand $h$ appears in all of these partitions. Clearly it appears at least once in $p(n - h)$ partitions. It appears at least twice in $p(n - 2h)$ partitions. It appears at least three times in $p(n - 3h)$ partitions. Hence, the total number of appearances of $h$ is

$$p(n - h) + p(n - 2h) + p(n - 3h) + \cdots$$

Therefore,

$$np(n) = \sum_{h=1}^{n} h(p(n-h) + p(n-2h) + p(n-3h) + \cdots)$$

$$= \sum_{hk \leq n} hp(n-hk) = \sum_{j=1}^{n} p(n-j) \sum_{h|j} h = \sum_{j=1}^{n} p(n-j)\sigma(j),$$

as asserted in (6.17).

We shall now prove that for any positive number $\epsilon > 0$, there corresponds a (possibly large) positive constant $C = C(\epsilon)$ such that

$$1 \leq p(n) \leq C(1+\epsilon)^n. \qquad (6.18)$$

First we note that the infinite series

$$\sum_{j=1}^{\infty} \frac{j(j+1)}{2(1+\epsilon)^j}$$

is convergent. This follows immediately by the ratio test. Choose one integer $N = N(\epsilon)$ so that

$$N \geq \sum_{j=1}^{\infty} \frac{j(j+1)}{2(1+\epsilon)^j}.$$

Now take

$$C = \max_{1 \leq n \leq N} \frac{p(n)}{(1+\epsilon)^n}. \qquad (6.19)$$

We will prove by induction on $n$ that (6.18) is true with $C$ defined by (6.19). Clearly by the definition of $C$, Eq. (6.18) is true for $n \leq N$. Suppose now we know that (6.18) is true up to but not including a particular $n(> N)$. Then by (6.17),

$$p(n) = \frac{1}{n} \sum_{j=1}^{n} p(n-j)\sigma(j) \leq \frac{1}{n} \sum_{j=1}^{n} C(1+\epsilon)^{n-j}\sigma(j)$$

$$= C(1+\epsilon)^n \frac{1}{n} \sum_{j=1}^{n} \frac{\sigma(j)}{(1+\epsilon)^j} \leq C(1+\epsilon)^n \frac{1}{n} \sum_{j=1}^{n} \frac{(1+2+3+\cdots+j)}{(1+\epsilon)^j}$$

$$= C(1+\epsilon)^n \frac{1}{n} \sum_{j=1}^{n} \frac{j(j+1)}{2(1+\epsilon)^j} \leq C(1+\epsilon)^n \frac{1}{N} \sum_{j=1}^{\infty} \frac{j(j+1)}{2(1+\epsilon)^j}$$

$$\leq C(1+\epsilon)^n \qquad \text{(by our choice of } N\text{)}.$$

Hence, (6.18) is proved by mathematical induction. Therefore,

$$1 \leqq \liminf_{n\to\infty} p(n)^{1/n} \leqq \limsup_{n\to\infty} p(n)^{1/n} \leqq \lim_{n\to\infty} C^{1/n}(1+\epsilon) = 1+\epsilon.$$

But $\epsilon$ is an arbitrary positive number. Therefore,

$$\limsup_{n\to\infty} p(n)^{1/n} = 1.$$

Hence,

$$\lim_{n\to\infty} p(n)^{1/n} = 1,$$

as desired.

The convergence of $p(n)^{1/n}$ to 1 which we have just proved is clearly visible in the following table. For comparison, we also show the convergence of $F_n^{1/n}$ to the golden mean $(1+\sqrt{5})/2 = 1.618\ldots$ (recall that $F_n$ denotes the $n$th Fibonacci number).

| $n$ | $p(n)^{1/n}$ | $F_n^{1/n}$ |
|-----|--------------|-------------|
| 5 | 1.475 | 1.380 |
| 10 | 1.453 | 1.493 |
| 100 | 1.210 | 1.605 |
| 250 | 1.141 | 1.613 |
| 1000 | 1.075 | 1.617 |

The value of knowing this result is (among other things) that it allows us to conclude (via the root test, see Appendix A) that

$$\sum_{n=0}^{\infty} p(n)q^n$$

is an absolutely convergent infinite series for $|q| < 1$.

An alternative, elementary proof appears in Andrews (1971c).

# Chapter 7
## Gaussian polynomials

Our knowledge of partitions and generating functions will allow us to make useful generalizations of the well-known binomial numbers (a.k.a. binomial coefficients) and their various identities. We are led to polynomials in $q$ called Gaussian polynomials (a.k.a. $q$-binomial numbers or $q$-binomial coefficients).

---

### Highlights of this chapter

- The binomial theorem and the binomial series say that the binomial numbers $\binom{n}{k}$ appear as coefficients in $(1 + z)^n$ and $(1 - z)^{-n}$, respectively.
- A $q$-analog of a mathematical object is some polynomial in $q$ such that the original object is retrieved when $q$ is set to 1. Such a $q$-analog of the binomial numbers are the $q$-binomial numbers, which can be defined by counting the Ferrers boards that fit inside an $N$-by-$m$ box.
- The $q$-binomial numbers are also called *Gaussian polynomials*. We present several identities and limits for Gaussian polynomials, which will be useful in later chapters.

---

## 7.1 Properties of the binomial numbers

You are probably acquainted with the *binomial numbers*: $\binom{n}{j}$ is defined combinatorially as the number of ways to choose a subset of $j$ elements from a set of $n$ elements. It is easy to show that the binomial numbers have the following simple, explicit formula:

$$\binom{n}{j} = \frac{n!}{j!(n-j)!} \quad \frac{n(n-1)(n-2)\cdots(n-j+1)}{j(j-1)(j-2)\cdots 1} \quad \text{for } n \ge j \ge 0.$$

The binomial numbers are usually presented in a triangular table called *Pascal's triangle*:

$$\binom{0}{0}$$
$$\binom{1}{0} \qquad \binom{1}{1}$$
$$\binom{2}{0} \qquad \binom{2}{1} \qquad \binom{2}{2}$$
$$\binom{3}{0} \qquad \binom{3}{1} \qquad \binom{3}{2} \qquad \binom{3}{3}$$
$$\binom{4}{0} \qquad \binom{4}{1} \qquad \binom{4}{2} \qquad \binom{4}{3} \qquad \binom{4}{4}$$
$$\vdots \qquad\qquad \vdots \qquad\qquad \vdots$$

or with the actual numbers:

$$1$$
$$1 \qquad 1$$
$$1 \qquad 2 \qquad 1$$
$$1 \qquad 3 \qquad 3 \qquad 1$$
$$1 \qquad 4 \qquad 6 \qquad 4 \qquad 1$$
$$\vdots \qquad\qquad \vdots \qquad\qquad \vdots$$

Counting the top row as the zeroth row, you will notice that the second and third rows contain the coefficients of $(1+z)^2 = 1 + 2z + z^2$ and $(1+z)^3 = 1 + 3z + 3z^2 + z^3$, respectively. Indeed, the binomial numbers on the $n$th row are the coefficients of $(1+z)^n$, as asserted by the celebrated *binomial theorem*.

**Theorem 6 (binomial theorem)**

$$(1+z)^n = \sum_{j=0}^{n} \binom{n}{j} z^j.$$

For negative powers, there is instead the *binomial series*, a result we utilized for some of our computations in Chapter 6.

**Theorem 7 (binomial series)**

$$(1-z)^{-n} = \sum_{j=0}^{\infty} \binom{n+j-1}{j} z^j,$$

*for $|z| < 1$.*

The binomial numbers have many nice and significant properties. Among these
are symmetry:

$$\binom{n+j}{j} = \binom{n+j}{n};$$

a nice formula for the sum of a row of binomial numbers:

$$\sum_{j=0}^{n} \binom{n}{j} = 2^n;$$

an even simpler formula for the alternating sum:

$$\sum_{j=0}^{n} (-1)^j \binom{n}{j} = 0 \quad \text{for } n \geq 1;$$

and a recurrence:

$$\binom{n}{j} = \binom{n-1}{j} + \binom{n-1}{j-1} \quad \text{for } n > j > 0,$$

with the initial values $\binom{n}{0} = \binom{n}{n} = 1$. You are invited to prove all these properties
in the following exercises.

---

## EXERCISES

89. Prove the formula $\binom{n}{j} = \frac{n!}{j!(n-j)!}$ from the combinatorial definition. (Difficulty rating: 2)

90. Prove the symmetric property of the binomial numbers in two ways, both
algebraically and combinatorially. (Difficulty rating: 2)

91. Use the recurrence to compute the fifth row of Pascal's triangle. (Difficulty
rating: 1)

92. Prove the recurrence in two ways, both algebraically and combinatorially.
(Difficulty rating: 2)

93. Prove the binomial theorem, combinatorially as well as by induction. (Difficulty rating: 2)

94. Use the binomial theorem to prove the two sum formulas by setting $z = 1$
and $z = -1$, respectively. (Difficulty rating: 1)

95. Prove the binomial series by induction. (Hint: $(1 - z)^{-(n-1)} = (1 - z)(1 -
z)^{-n}$. Use the recurrence.) (Difficulty rating: 2)

---

## 7.2 Lattice paths and the $q$-binomial numbers

Suppose we consider the following question. How many paths are there from the point $(0, 0)$ in the plane to the point $(2, 2)$ in which the only steps in the path are unit steps either vertically upward or horizontally to the right? Such paths are called *lattice paths*, and in this case the answer is six:

The explanation to the answer six is that $6 = \binom{2+2}{2}$, the number of ways to choose two horizontal steps out of a total of four steps (the remaining two being vertical). By the same argument, the number of such lattice paths from $(0, 0)$ to $(N, m)$ is given by $\binom{N+m}{m}$.

Now let us make a refinement of this result by inserting squares above-left of the paths:

If we regard these $\binom{2+2}{2}$ figures as Ferrers boards of partitions, they correspond to the six Ferrers boards that fit in a two-by-two box, that is, partitions into at most two parts, each part at most of size two:

$$\emptyset \quad 1 \quad 1+1 \quad 2 \quad 2+1 \quad 2+2$$

In analogy with the combinatorial definition of the binomial numbers, we now define a $q$-*binomial number* to be the generating function (in the variable $q$) for these $\binom{2+2}{2}$ partitions:

$$\begin{bmatrix} 2+2 \\ 2 \end{bmatrix} = q^0 + q^1 + q^{1+1} + q^2 + q^{2+1} + q^{2+2} = q^0 + q^1 + 2q^2 + q^3 + q^4.$$

More generally, the $q$-binomial numbers are defined

$$\begin{bmatrix} N+m \\ m \end{bmatrix} = \sum_{n \geq 0} p(n \mid \leq m \text{ parts, each} \leq N)q^n.$$

This is a so called $q$-*analog* of the binomial numbers, which means that it is a natural refinement such that for $q = 1$, we retrieve $\binom{N+m}{m}$. We shall see that the many properties of binomial numbers also carry over to the $q$-analog. For example, the symmetric property

$$\begin{bmatrix} N+m \\ m \end{bmatrix} = \begin{bmatrix} N+m \\ N \end{bmatrix}$$

follows immediately by conjugation of the Ferrers boards.

Then there is the recurrence for binomial numbers, which we may write as

$$\binom{N+m}{m} = \binom{N+m-1}{m-1} + \binom{N+m-1}{N-1}.$$

In terms of Ferrers boards, this recurrence says that the set of boards that fit in an $N$-by-$m$ box can be partitioned into two disjoint sets: boards that actually fit into an $N$-by-$(m-1)$ box and boards that don't. In the latter case, these Ferrers boards have a first column of length $m$ which upon removal leaves a Ferrers board that fits into an $(N-1)$–by–$m$ box.

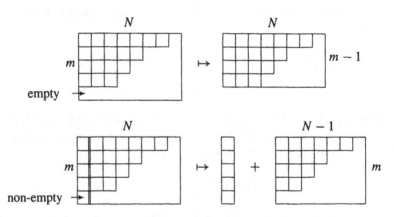

Refining to partitions of $n$, the same argument can be expressed as

$$p(n \mid \leq m \text{ parts, each} \leq N)q^n$$
$$= p(n \mid \leq m-1 \text{ parts, each} \leq N)q^n$$
$$+ q^m p(n-m \mid \leq m \text{ parts, each} \leq N-1)q^{n-m}.$$

Summation over $n$ proves the $q$-analog recurrence:

$$\begin{bmatrix} N+m \\ m \end{bmatrix} = \begin{bmatrix} N+m-1 \\ m-1 \end{bmatrix} + q^m \begin{bmatrix} N+m-1 \\ N-1 \end{bmatrix}. \tag{7.1}$$

Observe that by conjugation of the same argument, we also obtain an alternative recurrence:

$$\begin{bmatrix} N+m \\ m \end{bmatrix} = q^N \begin{bmatrix} N+m-1 \\ m-1 \end{bmatrix} + \begin{bmatrix} N+m-1 \\ N-1 \end{bmatrix}. \tag{7.2}$$

Next we have the explicit formula for the binomial numbers,

$$\binom{N}{m} = \frac{N(N-1)(N-2)\cdots(N-m+1)}{m(m-1)(m-2)\cdots 1}.$$

The analogous formula for $q$-binomial numbers looks like

$$\begin{bmatrix} N \\ m \end{bmatrix} = \frac{(1-q^N)(1-q^{N-1})\cdots(1-q^{N-m+1})}{(1-q^m)(1-q^{m-1})\cdots(1-q)}.$$

This is perhaps most easily remembered as the $q$-analog working on every integer factor $i$ and replacing it with the polynomial $(1+q+\cdots+q^{i-1}) = (1-q^i)/(1-q)$. The proof is left as an exercise.

---

**EXERCISES**

96. Explain, combinatorially, why the $q$-binomial numbers satisfy the initial values $\begin{bmatrix} N \\ 0 \end{bmatrix} = \begin{bmatrix} N \\ N \end{bmatrix}$. (Difficulty rating: 1)

97. Prove the formula for $q$-binomial numbers by induction, using the recurrence for the induction step. (Difficulty rating: 2)

---

## 7.3 The $q$-binomial theorem and the $q$-binomial series

Our objective now is to find $q$-analogs of the binomial theorem and the binomial series. Since the $q$-binomial numbers are generating functions for certain partitions, the trick must be to enter the variable $q$ into the left-hand side expressions $(1+z)^n$ and $(1-z)^{-n}$ of the binomial theorem and binomial series, respectively, so that also here partitions are also counted.

How this can be done has already been explained in Chapter 5. Equations (5.9) and (5.11) specialize to the following identities:

$$\prod_{j=1}^{N}(1+zq^j) = \sum_{n=0}^{\infty}\sum_{m=0}^{\infty} p(n \mid m \text{ distinct parts each} \leqq N)z^m q^n \quad (7.3)$$

and

$$\prod_{j=1}^{N}\frac{1}{1-zq^j} = \sum_{n=0}^{\infty}\sum_{m=0}^{\infty} p(n \mid m \text{ parts each} \leqq N)z^m q^n. \quad (7.4)$$

To begin with the $q$-binomial theorem, it seems very reasonable to model its left-hand side on (7.3). But the partitions in (7.3) have $m$ distinct parts each $\leq N$, whereas the partitions counted by $\begin{bmatrix} N \\ m \end{bmatrix}$ have at most $m$ parts, not necessarily distinct, each $\leq N - m$. However, there is a simple bijection between these two sets of partitions: from the former partitions, remove $i$ from the $i$th smallest part for all $i$ from 1 to $m$:

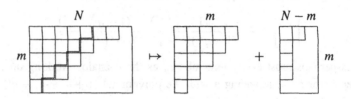

In the bijection we have removed $1 + 2 + \cdots + m = m(m + 1)/2$ squares, so the theorem takes the following form:

**Theorem 8** ($q$-**binomial theorem**)

$$\prod_{j=1}^{N}(1 + zq^{j}) = \sum_{m=0}^{\infty} q^{m(m+1)/2} \begin{bmatrix} N \\ m \end{bmatrix} z^{m}.$$

For the $q$-binomial series, we run into the same kind of problem, with an even simpler solution. If we model its left-hand side on (7.4), we must transform partitions with $m$ parts to partitions with at most $m$ parts. The obvious solution is to remove the first column (of length $m$) of the Ferrers board. Since the first partitions have $m$ parts each $\leq N$, the transformed partitions will have at most $m$ parts each $\leq N - 1$.

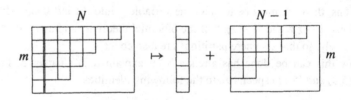

This is a bijection proving the $q$-binomial series:

**Theorem 9** ($q$-**binomial series**)

$$\prod_{j=1}^{N} \frac{1}{1 - zq^{j}} = \sum_{m=0}^{\infty} q^{m} \begin{bmatrix} N + m - 1 \\ m \end{bmatrix} z^{m},$$

*for $|z| < 1$ and $|q| < 1$.*

**EXERCISE**

98. Derive two identities by setting $z = 1$ and $z = -1$, respectively, in the $q$-binomial theorem. (Difficulty rating: 1)

## 7.4 Gaussian polynomial identities

The $q$-binomial numbers are also called *Gaussian polynomials*. They are polynomials in $q$ by the combinatorial definition, and it is clear from the formula

$$\begin{bmatrix} N \\ m \end{bmatrix} = \frac{(1 - q^N)(1 - q^{N-1}) \cdots (1 - q^{N-m+1})}{(1 - q^m)(1 - q^{m-1}) \cdots (1 - q)}$$

that the degree must be $mN - m(m-1)/2 - m(m+1)/2 = mN - m^2 = m(N - m)$.

Gauss was perhaps not the first person to define these polynomials, but he did prove the following formula in order to settle the sign of the Gaussian sum.

**Theorem 10 (Gaussian formula)**

$$\sum_{j=0}^{n} (-1)^j \begin{bmatrix} n \\ j \end{bmatrix} = \begin{cases} 0 & \text{if } n \text{ is odd} \\ (1 - q)(1 - q^3)(1 - q^5) \cdots (1 - q^{n-1}) & \text{if } n \text{ is even.} \end{cases}$$

The case when $n$ is odd is, in fact, not so subtle after all, because if $n$ is odd, then

$$\sum_{j=0}^{n} (-1)^j \begin{bmatrix} n \\ j \end{bmatrix} = \sum_{j=0}^{n} (-1)^{n-j} \begin{bmatrix} n \\ n-j \end{bmatrix}$$

$$= (-1)^n \sum_{j=0}^{n} (-1)^j \begin{bmatrix} n \\ j \end{bmatrix} \qquad \text{(by symmetry)}$$

$$= -\sum_{j=0}^{n} (-1)^j \begin{bmatrix} n \\ j \end{bmatrix}, \qquad (7.5)$$

but a polynomial can only equal its negative if it is 0. So the top line of the Gaussian formula is true.

Suppose that we denote the left-hand side of the Gaussian formula by $f(n)$. Then, by the recurrence (7.1) for $q$-binomial numbers,

$$f(n) = \sum_{j=0}^{n} (-1)^j \left( \begin{bmatrix} n-1 \\ j \end{bmatrix} + q^{n-j} \begin{bmatrix} n-1 \\ j-1 \end{bmatrix} \right)$$

$$= f(n-1) + (-1)^n \sum_{j=0}^{n} (-1)^j q^j \begin{bmatrix} n-1 \\ j \end{bmatrix},$$

whereas the alternative recurrence (7.2) yields

$$f(n) = \sum_{j=0}^{n} (-1)^j \left( \begin{bmatrix} n-1 \\ j-1 \end{bmatrix} + q^{n-j} \begin{bmatrix} n-1 \\ j \end{bmatrix} \right)$$

$$= -f(n-1) + \sum_{j=0}^{n} (-1)^j q^j \begin{bmatrix} n-1 \\ j \end{bmatrix}.$$

Adding the two above expressions for $f(n)$, we find that if $n$ is even, then

$$f(n) = \sum_{j=0}^{n} (-1)^j q^j \begin{bmatrix} n-1 \\ j \end{bmatrix}$$

$$= \sum_{j=0}^{n} (-1)^j \left( 1 - (1 - q^j) \right) \begin{bmatrix} n-1 \\ j \end{bmatrix}$$

$$= -\sum_{j=0}^{n} (-1)^j (1 - q^j) \begin{bmatrix} n-1 \\ j \end{bmatrix} \qquad \text{(by the top line of the Gaussian formula)}$$

$$= -(1 - q^{n-1}) \sum_{j=0}^{n} (-1)^j \begin{bmatrix} n-2 \\ j-1 \end{bmatrix}$$

$$= (1 - q^{n-1}) f(n-2).$$

Hence, when $n$ is even,

$$f(n) = (1 - q^{n-1}) f(n-2)$$
$$= (1 - q^{n-1})(1 - q^{n-3}) f(n-4)$$
$$= (1 - q^{n-1})(1 - q^{n-3}) \cdots (1 - q^3)(1 - q) f(0)$$
$$= (1 - q^{n-1})(1 - q^{n-3}) \cdots (1 - q^3)(1 - q)$$

as desired, and the Gaussian formula is proved.

There is also the old chestnut concerning the sum of the squares of the binomial numbers:

$$\sum_{j=0}^{n}\binom{n}{j}^2 = \binom{2n}{n}.$$

Rather than recount directly the proof of this classic identity, we shall show the $q$-analog

$$\sum_{j=0}^{n}q^{j^2}\begin{bmatrix}n\\j\end{bmatrix}^2 = \begin{bmatrix}2n\\n\end{bmatrix}. \tag{7.6}$$

In fact, this is even easier than the proof of the Gaussian formula, namely by theorem 8,

$$\sum_{j=0}^{2n}z^j q^{\frac{j(j+1)}{2}}\begin{bmatrix}2n\\j\end{bmatrix} = \prod_{j=1}^{n}(1+zq^j) = \prod_{j=1}^{n}(1+zq^j)\prod_{j=1}^{n}(1+(zq^n)q^j)$$

$$= \sum_{j=0}^{n}z^j q^{j(j+1)/2}\begin{bmatrix}n\\j\end{bmatrix}\sum_{k=0}^{n}(zq^n)^k q^{k(k+1)/2}\begin{bmatrix}n\\k\end{bmatrix}$$

(by two applications of the $q$-binomial theorem).

Comparing coefficients of $z^n$ on each side above, we see that

$$q^{n(n+1)/2}\begin{bmatrix}2n\\n\end{bmatrix} = \sum_{k=0}^{n}q^{(n-k)(n-k+1)/2}\begin{bmatrix}n\\n-k\end{bmatrix}q^{nk+k(k+1)/2}\begin{bmatrix}n\\k\end{bmatrix}$$

$$= \sum_{k=0}^{n}q^{n(n+1)/2+k^2}\begin{bmatrix}n\\k\end{bmatrix}^2,$$

where we used the symmetry of $q$-binomial numbers for the last step. Cancelling $q^{n(n+1)/2}$, we see that we have proved (7.6). We shall see in the next chapter that this identity is important.

---

**EXERCISES**

99. Prove that

$$\sum_{j=0}^{n}q^{\frac{j}{2}}\begin{bmatrix}n\\j\end{bmatrix} = (1+q^{\frac12})(1+q)(1+q^{\frac32})\cdots(1+q^{\frac{n}{2}}).$$

(Difficulty rating: 3)

100. Prove that

$$\sum_{j=0}^{n} q^j \begin{bmatrix} m+j \\ m \end{bmatrix} = \begin{bmatrix} n+m+1 \\ n \end{bmatrix}.$$

(Difficulty rating: 2)

---

## 7.5  Limiting values of Gaussian polynomials

In the next chapter, we shall need to know what happens to Gaussian polynomials as some (or maybe all) of the parameters get large.

The formula for Gaussian polynomials makes this an easy problem. First, for fixed $m$,

$$\lim_{N \to \infty} \begin{bmatrix} N \\ m \end{bmatrix} = \lim_{N \to \infty} \frac{\prod_{j=1}^{N}(1-q^j)}{\prod_{j=1}^{m}(1-q^j)\prod_{j=1}^{N-m}(1-q^j)}$$

$$= \frac{\prod_{j=1}^{\infty}(1-q^j)}{\prod_{j=1}^{m}(1-q^j)\prod_{j=1}^{\infty}(1-q^j)} = \frac{1}{\prod_{j=1}^{m}(1-q^j)}. \quad (7.7)$$

Next, for fixed $m_1$ and $m_2$, with $R > S$ positive,

$$\lim_{N \to \infty} \begin{bmatrix} RN+m_1 \\ SN+m_2 \end{bmatrix} = \lim_{N \to \infty} \frac{\prod_{j=1}^{RN+m_1}(1-q^j)}{\prod_{j=1}^{SN+m_2}(1-q^j)\prod_{j=1}^{(R-S)N+m_1-m_2}(1-q^j)}$$

$$= \frac{\prod_{j=1}^{\infty}(1-q^j)}{\prod_{j=1}^{\infty}(1-q^j)\prod_{j=1}^{\infty}(1-q^j)} = \frac{1}{\prod_{j=1}^{\infty}(1-q^j)}. \quad (7.8)$$

# Chapter 8

## Durfee squares

In Eq. (7.6) in the previous chapter, we proved the Gaussian polynomial identity

$$\sum_{j=0}^{n} q^{j^2} \begin{bmatrix} n \\ j \end{bmatrix}^2 = \begin{bmatrix} 2n \\ n \end{bmatrix}. \qquad (8.1)$$

Now it is possible to understand this identity purely by means of a new classification of partitions according to something called the *Durfee square*.

---

**Highlights of this chapter**

- The Ferrers board of a partition can be decomposed into two smaller boards joined by a square shape called the Durfee square.
- Related to the Durfee square is a representation of partitions called the Frobenius symbol.
- The generating function for Frobenius symbols gives a simple proof of Jacobi's triple product identity, a famous and very useful theorem.
- Using Gaussian polynomials and Jacobi's triple product, we prove the Rogers-Ramanujan identities.
- Decomposition of partitions into successive Durfee squares yields a nice generalization of the first Rogers-Ramanujan identity.

---

## 8.1 Durfee squares and generating functions

Let us recall the Ferrers board of a partition. For example, the Ferrers board of $4 + 4 + 2 + 1 + 1$ is

We readily see a largest possible square contained within the Ferrers board and anchored in the upper left-hand corner of the Ferrers board. This is called the *Durfee square* of the partition. In this instance, the partition $4 + 4 + 2 + 1 + 1$ has a Durfee square of side 2.

If you prefer a more operational definition, you may say that a given partition has a Durfee square of side $s$ if the $s$th part (numbering from largest to smallest) is $\geq s$ but the $(s + 1)$st part is $\leq s$.

From this perspective, let us reconsider (8.1). From Chapter 7, we know that the $q$-binomial number

$$\begin{bmatrix} 2n \\ n \end{bmatrix}$$

is the generating function for all partitions with at most $n$ parts, each $\leq n$.

Of these partitions, what is the generating function for partitions with Durfee square of side $j$? We can answer this fairly easily by the following generic representation of the Ferrers board of such partitions:

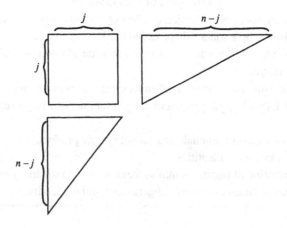

The partition made up of the $j \times j$ square is obviously generated by the monomial

$$q^{j \times j} = q^{j^2}.$$

By the definition of $q$-binomial numbers in Chapter 7, the upper right portion is generated by

$$\begin{bmatrix} n \\ j \end{bmatrix},$$

and the lower left portion is produced by

$$\begin{bmatrix} n \\ n - j \end{bmatrix}.$$

Hence, the generating function for partitions into at most $n$ parts, each $\leqq n$, with Durfee square of side $j$ is

$$q^{j^2} \begin{bmatrix} n \\ j \end{bmatrix} \begin{bmatrix} n \\ n - j \end{bmatrix} = q^{j^2} \begin{bmatrix} n \\ j \end{bmatrix}^2.$$

But every partition enumerated by $\begin{bmatrix} 2n \\ n \end{bmatrix}$ has a unique Durfee square with $0 \leqq j \leqq n$. Therefore,

$$\sum_{j=0}^{n} q^{j^2} \begin{bmatrix} n \\ j \end{bmatrix}^2 = \begin{bmatrix} 2n \\ n \end{bmatrix},$$

which is (8.1).

If we let $n \to \infty$, we may directly deduce a formula originally found by Jacobi:

$$\sum_{j=0}^{\infty} \frac{q^{j^2}}{(1 - q)^2 (1 - q^2)^2 \cdots (1 - q^j)^2} = \prod_{n=1}^{\infty} \frac{1}{1 - q^n}. \qquad (8.2)$$

---

## EXERCISES

101. Examine all partitions with parts $\leq N$ while noting both the size of the Durfee square and the number of parts. Prove that

$$\sum_{j=0}^{N} \begin{bmatrix} N \\ j \end{bmatrix} \frac{z^j q^{j^2}}{(1 - zq)(1 - zq^2) \cdots (1 - zq^j)} = \prod_{n=1}^{N} \frac{1}{1 - zq^n}.$$

(Difficulty rating: 2)

102. Show that the generating function for self-conjugate partitions with each part $\leq N$ is

$$\sum_{j=0}^{N} q^{j^2} \begin{bmatrix} N \\ j \end{bmatrix}_{q^2},$$

where $\begin{bmatrix} N \\ j \end{bmatrix}_{q^2}$ is the Gaussian polynomial $\begin{bmatrix} N \\ j \end{bmatrix}$, with $q$ replaced by $q^2$. (Difficulty rating: 2)

103. Deduce from the formula for Gaussian polynomials of Chapter 7 that

$$\sum_{j=0}^{N} q^{j^2} \begin{bmatrix} N \\ j \end{bmatrix}_{q^2} = (1+q)(1+q^3)\cdots(1+q^{2N-1}).$$

(Difficulty rating: 2)

104. Conclude from Exercises 102 and 103 that the number of partitions of $n$ into distinct odd parts, each $\leq 2N - 1$, equals the number of self-conjugate partitions of $n$ whose parts are each $\leq N$. (Difficulty rating: 2)

105. Construct a bijective proof of the assertion in Exercise 104. (Difficulty rating: 3)

---

## 8.2 Frobenius symbols

From the Ferrers board of a partition, we may construct an entirely new numerical representation of a partition that immediately reveals the size of the Durfee square and the conjugate partition. This new representation is called the *Frobenius symbol* of the partition, and it is constructed as follows for $7 + 7 + 4 + 2 + 2$:

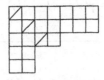

The symbol consists of two rows of decreasing nonnegative integers. The rows are each of length $s$, where $s$ is the size of the Durfee square. The $j$th entry on the top row consists of the number of boxes on the $j$th row of the Ferrers board to the right of the diagonal (shaded in the diagram). The $j$th entry on the bottom row consists of the number of boxes in the $j$th column of the Ferrers board below the diagonal.

So the Frobenius symbol for the partition $7 + 7 + 4 + 2 + 2$ is

$$\begin{pmatrix} 6 & 5 & 1 \\ 4 & 3 & 0 \end{pmatrix},$$

and clearly from this symbol we can reconstruct the original partition.

It is now a fairly straightforward matter to generate partitions represented by Frobenius symbols. Namely, we examine the coefficient of $z^0$ in

$$\left\{(1 + (zq)q^0)(1 + (zq)q^1)(1 + (zq)q^2)(1 + (zq)q^3)\cdots\right\}$$
$$\times \left\{(1 + z^{-1}q^0)(1 + z^{-1}q^1)(1 + z^{-1}q^2)(1 + z^{-1}q^3)\cdots\right\}. \quad (8.3)$$

When we multiply out these products, we find that to get $z^0$, we need exactly the same number, say $s$, of second terms selected from each product contained in curly brackets. So a typical term will look like

$$q^s \cdot q^{a_1+a_2+\cdots+a_s} q^{b_1+b_2+\cdots+b_s},$$

where $a_1 > a_2 > \cdots > a_s \geqq 0$ and $b_1 > b_2 > \cdots > b_s \geqq 0$. But this corresponds to the Frobenius symbol

$$\begin{pmatrix} a_1 a_2 \cdots a_s \\ b_1 b_2 \cdots b_s \end{pmatrix}$$

related to a unique partition of $s + \sum a_i + \sum b_i$.

Since there is a perfect correspondence of the Frobenius symbols to the partitions of $n$, we may conclude that the coefficient of $z^0$ in (8.3) is

$$\sum_{n=0}^{\infty} p(n)q^n = \prod_{n=1}^{\infty} \frac{1}{1 - q^n}. \quad (8.4)$$

## 8.3 Jacobi's triple product identity

The seemingly natural view of the generating function for Frobenius symbols allows us to prove easily a famous theorem of Jacobi. This idea has also been applied in a much more general setting with extensive applications (Andrews, 1984).

**Theorem 11 (Jacobi's triple product identity)**

$$\sum_{n=-\infty}^{\infty} z^n q^{\frac{n(n+1)}{2}} = \prod_{n=1}^{\infty}(1 - q^n)(1 + zq^n)(1 + z^{-1}q^{n-1}).$$ (8.5)

*for* $|q| < 1$, $z \neq 0$.

In order to prove Jacobi's triple product identity, we define

$$J(z) = \prod_{n=1}^{\infty}(1 + zq^n)(1 + z^{-1}q^{n-1}).$$

We may expand $J(z)$ in a Laurent series around $z = 0$, so

$$J(z) = \sum_{n=-\infty}^{\infty} A_n(q)z^n.$$ (8.6)

Furthermore,

$$J(zq) = \prod_{n=1}^{\infty}(1 + zq^{n+1})(1 + z^{-1}q^{n-2})$$

$$= (1 + z^{-1}q^{-1})\prod_{n=2}^{\infty}(1 + zq^n)\prod_{n=1}^{\infty}(1 + z^{-1}q^{n-1})$$

$$= z^{-1}q^{-1}\prod_{n=1}^{\infty}(1 + zq^n)\prod_{n=1}^{\infty}(1 + z^{-1}q^{n-1})$$

$$= z^{-1}q^{-1}J(z).$$

Comparing coefficients of $z^n$ on both sides, we see that

$$q^n A_n(q) = q^{-1}A_{n+1}(q).$$

Iteration of this recursion reveals that for all $n$,

$$A_n(q) = q^{\frac{n(n+1)}{2}} A_0(q).$$ (8.7)

But by Eq. (8.4) and the comments preceding it,

$$A_0(q) = \prod_{n=1}^{\infty}\frac{1}{1 - q^n},$$ (8.8)

Combining (8.8), (8.7), and (8.6), we deduce

$$\prod_{n=1}^{\infty}(1 + zq^n)(1 + z^{-1}q^{n-1}) = J(z) = A_0(q) \sum_{n=-\infty}^{\infty} z^n q^{\frac{n(n+1)}{2}}$$

$$= \prod_{n=1}^{\infty} \frac{1}{1 - q^n} \sum_{n=-\infty}^{\infty} z^n q^{\frac{n(n+1)}{2}},$$

which is clearly equivalent to Jacobi's triple product identity.

---

**EXERCISES**

106. By replacing $q$ by $q^3$ and $z$ by $-q^{-1}$, deduce Euler's pentagonal number theorem from Jacobi's triple product. (Difficulty rating: 1)

107. Prove the identity $\sum_{n=-\infty}^{\infty}(-1)^n q^{n^2} = \prod_{n=1}^{\infty}\frac{(1-q^n)}{(1+q^n)}$. (Difficulty rating: 2)

108. Prove the identity $\sum_{n=0}^{\infty} q^{n(n+1)/2} = \prod_{n=1}^{\infty}\frac{(1-q^{2n})}{(1-q^{2n-1})}$. (Difficulty rating: 2)

109. Prove Jacobi's identity $\sum_{n=0}^{\infty} (-1)^n q^{n(n+1)/2}(2n + 1) = \prod_{n=1}^{\infty}(1 - q^n)^3$. (Difficulty rating: 3)

---

## 8.4 The Rogers-Ramanujan identities

There is no really easy proof of the Rogers-Ramanujan identities. However, our study of Gaussian polynomials and Jacobi's triple product puts us in a position to prove some polynomial identities of David Bressoud (1981), which in turn will yield the Rogers-Ramanujan identities.

Following Chapman (2002), we shall take things step by step. First we consider five sequences of polynomials:

$$s_n(q) = \sum_{j=0}^{n} q^{j^2} \begin{bmatrix} n \\ j \end{bmatrix}, \tag{8.9}$$

$$t_n(q) = \sum_{j=0}^{n} q^{j^2+j} \begin{bmatrix} n \\ j \end{bmatrix}, \tag{8.10}$$

$$\sigma_n(q) = \sum_{j=-\infty}^{\infty} (-1)^j q^{j(5j+1)/2} \begin{bmatrix} 2n \\ n + 2j \end{bmatrix}, \tag{8.11}$$

$$\sigma_n^*(q) = \sum_{j=-\infty}^{\infty} (-1)^j q^{j(5j+1)/2} \begin{bmatrix} 2n + 1 \\ n + 1 + 2j \end{bmatrix}, \tag{8.12}$$

and

$$\tau_n(q) = \sum_{j=-\infty}^{\infty} (-1)^j q^{j(5j-3)/2} \begin{bmatrix} 2n+1 \\ n+2j \end{bmatrix}. \tag{8.13}$$

In the following pages, we shall prove using mathematical induction that

$$s_n(q) = \sigma_n(q) = \sigma_n^*(q) \tag{8.14}$$

and

$$t_n(q) = \tau_n(q).$$

Once this has been achieved, it will be a straightforward matter to show that

$$1 + \sum_{j=1}^{\infty} \frac{q^{j^2}}{(1-q)(1-q^2)\cdots(1-q^j)} = \lim_{n\to\infty} s_n(q) = \lim_{n\to\infty} \sigma_n(q)$$

$$= \prod_{n=0}^{\infty} \frac{1}{(1-q^{5n+1})(1-q^{5n+4})}$$

and

$$1 + \sum_{j=1}^{\infty} \frac{q^{j^2+j}}{(1-q)(1-q^2)\cdots(1-q^j)} = \lim_{n\to\infty} t_n(q) = \lim_{n\to\infty} \tau_n(q)$$

$$= \prod_{n=0}^{\infty} \frac{1}{(1-q^{5n+2})(1-q^{5n+3})}.$$

Indeed it follows immediately from (7.7) of Chapter 7, that

$$\lim_{n\to\infty} s_n(q) = 1 + \sum_{j=1}^{\infty} \frac{q^{j^2}}{(1-q)(1-q^2)\cdots(1-q^j)}$$

and

$$\lim_{n\to\infty} t_n(q) = 1 + \sum_{j=1}^{\infty} \frac{q^{j^2+j}}{(1-q)(1-q^2)\cdots(1-q^j)}. \tag{8.15}$$

Thus we are treating polynomials that, in the limit as $n \to \infty$, converge to the left-hand sides of the Rogers-Ramanujan identities.

Now we are going to prove two recurrences for $s_n(q)$ and $t_n(q)$. First,

$$s_n(q) = \sum_{j\geq 0} q^{j^2} \left( \begin{bmatrix} n-1 \\ j \end{bmatrix} + q^{n-j} \begin{bmatrix} n-1 \\ j-1 \end{bmatrix} \right)$$

$$= s_{n-1}(q) + q^n \sum_{j\geq 0} q^{j^2+j} \begin{bmatrix} n-1 \\ j \end{bmatrix}$$

$$= s_{n-1}(q) + q^n t_{n-1}(q). \tag{8.16}$$

Second,

$$t_n(q) - q^n s_n(q) = \sum_{j \geq 0} q^{j^2+j} \begin{bmatrix} n \\ j \end{bmatrix} (1 - q^{n-j})$$

$$= (1 - q^n) \sum_{j \geq 0} q^{j^2+j} \begin{bmatrix} n-1 \\ j \end{bmatrix}$$

$$= (1 - q^n) t_{n-1}(q).$$

It is now a simple problem in mathematical induction to show that (8.13) and (8.14) plus the initial values $s_0(q) = t_0(q) = 1$ uniquely define $s_n(q)$ and $t_n(q)$ for all $n$.

We now want to show that $\sigma_n(q)$ and $\sigma_n^*(q)$ are identical. This is because

$$\sigma_n^*(q) = \sum_{j=-\infty}^{\infty} (-1)^j q^{j(5j+1)/2} \left( \begin{bmatrix} 2n \\ n+2j \end{bmatrix} + q^{n+1+2j} \begin{bmatrix} 2n \\ n+1+2j \end{bmatrix} \right)$$

$$= \sigma_n(q) + q^{n+1} \left( \sum_{j=0}^{\infty} (-1)^j q^{j(5j+5)/2} \begin{bmatrix} 2n \\ n+1+2j \end{bmatrix} \right.$$

$$\left. + \sum_{j=-\infty}^{-1} (-1)^j q^{j(5j+5)/2} \begin{bmatrix} 2n \\ n+1+2j \end{bmatrix} \right)$$

$$= \sigma_n(q) + q^{n+1} \left( \sum_{j=0}^{\infty} (-1)^j q^{j(5j+5)/2} \begin{bmatrix} 2n \\ n-1-2j \end{bmatrix} \right.$$

$$\left. + \sum_{j=0}^{\infty} (-1)^{-j-1} q^{(-j-1)(5(-j))/2} \begin{bmatrix} 2n \\ n-1-2j \end{bmatrix} \right)$$

$$= \sigma_n(q). \tag{8.17}$$

So in the following we will use either (8.11) or (8.12) to represent $\sigma_n(q)$. In particular,

$$\sigma_n(q) - \sigma_{n-1}(q)$$

$$= \sum_{j=-\infty}^{\infty} (-1)^j q^{j(5j+1)/2} \begin{bmatrix} 2n \\ n+2j \end{bmatrix} - \sum_{j=-\infty}^{\infty} (-1)^j q^{j(5j+1)/2} \begin{bmatrix} 2n-1 \\ n+2j \end{bmatrix}$$

$$= \sum_{j=-\infty}^{\infty} (-1)^j q^{j(5j+1)/2} \begin{bmatrix} 2n-1 \\ n+2j-1 \end{bmatrix} q^{n-2j} \qquad \text{(by (7.2))}$$

$$= q^n \tau_{n-1}(q). \tag{8.18}$$

Finally,

$$\tau_n(q) - q^n \sigma_n(q)$$

$$= \sum_{j=-\infty}^{\infty} (-1)^j q^{j(5j-3)/2} \left( \begin{bmatrix} 2n+1 \\ n+1-2j \end{bmatrix} - q^{n+2j} \begin{bmatrix} 2n \\ n-2j \end{bmatrix} \right)$$

$$= \sum_{j=-\infty}^{\infty} (-1)^j q^{j(5j-3)/2} \begin{bmatrix} 2n \\ n+1-2j \end{bmatrix} \qquad \text{(by (7.2))}$$

$$= \sum_{j=-\infty}^{\infty} (-1)^j q^{j(5j-3)/2} \left( \begin{bmatrix} 2n-1 \\ n-2j \end{bmatrix} + q^{n+1-2j} \begin{bmatrix} 2n-1 \\ n+1-2j \end{bmatrix} \right)$$

$$= \tau_{n-1}(q) + q^n \sum_{j=-\infty}^{\infty} (-1)^{1-j} q^{(1-j)(5(1-j)-3)/2 + 1 - 2(1-j)} \begin{bmatrix} 2n-1 \\ n-1+2j \end{bmatrix}$$

$$= \tau_{n-1}(q) - q^n \sum_{j=-\infty}^{\infty} (-1)^j q^{j(5j-3)/2} \begin{bmatrix} 2n-1 \\ n-1+2j \end{bmatrix}$$

$$= (1-q^n)\tau_{n-1}(q). \qquad\qquad (8.19)$$

But as we remarked earlier, the recurrences (8.15) and (8.16) plus $s_0(q) = t_0(q) = 1$ uniquely determine all the $s_n(q)$ and $t_n(q)$. Now we have just shown in (8.18) and (8.19) that $\sigma_n(q)$ and $\tau_n(q)$ satisfy precisely the same recurrences. Furthermore, $\sigma_0(q) = \tau_0(q) = 1$. Consequently for each $n \geqq 0$,

$$s_n(q) = \sigma_n(q)$$

and

$$t_n(q) = \tau_n(q).$$

Therefore, by (8.9),

$$1 + \sum_{j=1}^{\infty} \frac{q^{j^2}}{(1-q)(1-q^2)\cdots(1-q^j)}$$

$$= \lim_{n\to\infty} s_n(q) = \lim_{n\to\infty} \sigma_n(q)$$

$$= \sum_{j=-\infty}^{\infty} (-1)^j q^{j(5j+1)/2} \frac{1}{\prod_{m=1}^{\infty}(1-q^m)} \qquad \text{(by (7.8))}$$

$$= \frac{\prod_{m=1}^{\infty}(1-q^{5m})(1-q^{5m-2})(1-q^{5m-3})}{\prod_{m=1}^{\infty}(1-q^m)}$$

$$\text{(by Theorem 11 with } q \text{ replaced by } q^5 \text{ and } z \text{ by } -q^{-2})$$

$$= \prod_{m=1}^{\infty} \frac{1}{(1-q^{5m-4})(1-q^{5m-1})},$$

proving the first Rogers-Ramanujan identity. By (8.10),

$$1 + \sum_{j=1}^{\infty} \frac{q^{j^2+j}}{(1-q)(1-q^2)\cdots(1-q^j)}$$
$$= \lim_{n\to\infty} t_n(q) = \lim_{n\to\infty} \tau_n(q)$$
$$= \sum_{j=-\infty}^{\infty} (-1)^j q^{j(5j-3)/2} \frac{1}{\prod_{m=1}^{\infty}(1-q^m)} \qquad \text{(by (7.8))}$$
$$= \frac{\prod_{m=1}^{\infty}(1-q^{5m})(1-q^{5m-1})(1-q^{5m-4})}{\prod_{m=1}^{\infty}(1-q^m)}$$

(by Theorem 11 with $q$ replaced by $q^5$ and $z$ by $-q^{-1}$)

$$= \prod_{m=1}^{\infty} \frac{1}{(1-q^{5m-3})(1-q^{5m-2})},$$

proving the second Rogers-Ramanujan identity.

## 8.5 Successive Durfee squares

We shall here present a nice generalization of the first Rogers-Ramanujan identity. Let us reconsider the polynomials $s_n(q)$ defined in equation (8.10):

$$s_n(q) = \sum_{j=0}^{n} q^{j^2} \begin{bmatrix} n \\ j \end{bmatrix}.$$

Now we already know from Chapter 7 that $\begin{bmatrix} n \\ j \end{bmatrix}$ is the generating function for partitions with at most $j$ parts, each at most $n-j$. Noting that $j^2 = j + j + j + \cdots + j$, we see immediately that $s_n(q)$ is the generating function for partitions into parts each $\leqq n$, wherein each part is $\geqq$ the number of parts (our $j$ above).

For example,

$$s_4(q) = 1 + q^1 + q^2 + q^3 + q^4 + q^{2+2} + q^{3+2} + q^{4+2} + q^{3+3}$$
$$+ q^{4+3} + q^{4+4} + q^{3+3+3} + q^{4+3+3} + q^{4+4+3} + q^{4+4+4} + q^{4+4+4+4}.$$

If we examine such partitions with regard to their Durfee squares, we see that these partitions have no parts below their Durfee square.

With this beginning, it is possible to extend the concept of Durfee square to that of *successive Durfee squares*. The idea here is to find a succession of Durfee squares by examining the portion of the partition below a given square. For example, the partition

$$7+5+4+4+4+3+2+2+1+1$$

has five successive Durfee squares:

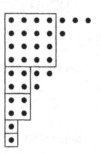

Once one considers the idea of successive Durfee squares, it is fairly easy to determine the generating function for partitions with at most $k$ successive Durfee squares. Indeed, we already know the answer when $k = 1$, namely,

$$s_n(q) = \sum_{j=0}^{n} q^{j^2} \begin{bmatrix} n \\ j \end{bmatrix}.$$

But for general $k$, we need only consider $k$ possible squares $j_1^2 \geq j_2^2 \geq j_3^2 \geq \cdots \geq j_k^2$ while noting how the interstices of the Ferrers board may be filled. This is most easily understood in the following schematic of the Ferrers board:

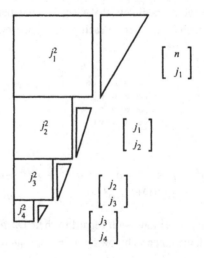

Hence, the generating function for partitions with at most $k$ successive Durfee squares, $s_{k,n}(q)$ is

$$s_{k,n}(q) = \sum_{n \geq j_1 \geq j_2 \geq \cdots \geq j_k \geq 0} q^{j_1^2 + j_2^2 + \cdots + j_k^2} \begin{bmatrix} n \\ j_1 \end{bmatrix} \begin{bmatrix} j_1 \\ j_2 \end{bmatrix} \begin{bmatrix} j_2 \\ j_3 \end{bmatrix} \cdots \begin{bmatrix} j_{k-1} \\ j_k \end{bmatrix}$$

$$= \sum_{n \geq j_1 \geq j_2 \geq \cdots \geq j_k \geq 0} \frac{q^{j_1^2 + j_2^2 + \cdots + j_k^2}(q;q)_n}{(q;q)_{n-j_1}(q;q)_{j_1-j_2}(q;q)_{j_2-j_3} \cdots (q;q)_{j_{k-1}-j_k}(q;q)_{j_k}},$$

where we have used the notation $(q;q)_m = (1-q)(1-q^2) \cdots (1-q^m)$. If we let $n \to \infty$, we find the complete generating function for all partitions with at most $k$ successive Durfee squares

$$s_{k,\infty}(q) = \sum_{j_1 \geq j_2 \geq \cdots \geq j_k \geq 0} \frac{q^{j_1^2 + j_2^2 + \cdots + j_k^2}}{(q;q)_{j_1-j_2}(q;q)_{j_2-j_3} \cdots (q;q)_{j_{k-1}-j_k}(q;q)_{j_k}}.$$

It is possible to prove (Andrews, 1979) that, in fact,

$$\sum_{j_1 \geq j_2 \geq \cdots \geq j_k \geq 0} \frac{q^{j_1^2 + j_2^2 + \cdots + j_k^2}}{(q;q)_{j_1-j_2}(q;q)_{j_2-j_3} \cdots (q;q)_{j_{k-1}-j_k}(q;q)_{j_k}}$$

$$= \prod_{\substack{n=1 \\ n \not\equiv 0, \pm(k+1) \pmod{2k+3}}}^{\infty} \frac{1}{(1-q^n)}. \tag{8.20}$$

Thus we may deduce that the number of partitions of $n$, each with at most $k$ successive Durfee squares, equals the number of partitions of $n$ into parts not congruent to $0, \pm(k+1) \pmod{2k+3}$.

When $k = 1$, this reduces precisely to the first Rogers-Ramanujan identity.

# Chapter 9

## Euler refined

In Chapters 1 and 2, we examined in detail Euler's simple and elegant theorem:

The number of partitions of an integer $n$ into distinct parts equals the number of partitions of $n$ into odd parts.

As we have seen, this theorem foreshadows a number of further results, the Rogers-Ramanujan identities being the most celebrated. In this chapter, we shall delve deeper into Euler's theorem (2.1), presenting two combinatorial variations and concluding with a consideration of a very recent refinement.

---

### Highlights of this chapter

- Sylvester refined Euler's theorem by considering the number of odd parts, and the number of sequences of consecutive distinct parts, respectively.
- Fine's refinement of Euler's theorem instead considers the largest of the odd parts and Dyson's rank of the partition into distinct parts.
- Bousquet-Mélou and Eriksson's refinement considers odd parts of bounded size and so-called "lecture hall partitions" into distinct parts.

---

## 9.1 Sylvester's refinement of Euler

In the late nineteenth century, the colorful mathematician J. J. Sylvester (1884) first discovered that there is more to Euler's theorem than meets the eye. The theorem we now consider appears in a gigantic paper, entitled *A Constructive Theory of Partitions in Three Acts, an Interact, and an Exodion.*

**Theorem 12 (Sylvester's refinement)** *The number of partitions of n using exactly k odd parts (each of which may be repeated) equals the number of* *partitions of n into k separate sequences of consecutive integers. (A sequence might have only one term.)*

For example, when $n = 15$ and $k = 3$, the eleven partitions in the first class are

$$11 + 3 + 1, \ 9 + 5 + 1, \ 9 + 3 + 1 + 1 + 1, \ 7 + 5 + 3,$$
$$7 + 5 + 1 + 1 + 1, \ 7 + 3 + 1 + 1 + 1 + 1 + 1, \ 7 + 3 + 3 + 1 + 1,$$
$$5 + 5 + 3 + 1 + 1, \ 5 + 3 + 3 + 3 + 1, \ 5 + 3 + 3 + 1 + 1 + 1 + 1,$$
$$5 + 3 + 1 + 1 + 1 + 1 + 1 + 1 + 1,$$

and the eleven partitions in the second class are

$$11 + 3 + 1, \ 10 + 4 + 1, \ 9 + 5 + 1, \ 9 + 4 + 2,$$
$$8 + 6 + 1, \ 8 + 5 + 2, \ 8 + 4 + 2 + 1, \ 7 + 5 + 3, \ 7 + 5 + 2 + 1,$$
$$7 + 4 + 3 + 1, \ 6 + 5 + 3 + 1.$$

We now prove this result by producing a variation on the Ferrers graph for partitions into odd parts. Instead of lining up rows of dots or boxes to represent parts by left justification, we justify them *centrally* as follows:

Partitions with odd parts can clearly be represented with centrally justified graphs. For example, the partition $13 + 13 + 13 + 11 + 9 + 3 + 3 + 1$ is thus represented:

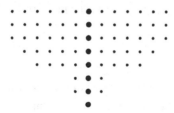

Now we take this representation and associate the dots in a new way:

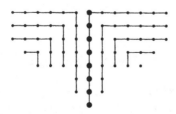

and the new partition is $14 + 12 + 11 + 8 + 7 + 6 + 4 + 3 + 1$.

We observe with some surprise that the new partition is one with distinct parts. With perhaps even more surprise we see that the first partition uses exactly five different odd numbers and the second partition consists of five separate sequences $14$, $12 + 11$, $8 + 7 + 6$, $4 + 3$, and $1$.

However, a little thought will convince us that this phenomenon does, in fact, always occur. The original graph is symmetric about the center. We see that each successive new right angle of dots (starting from the center) in the second diagram must, if it branches left, have at most the same number of dots vertically and at most one less dot horizontally as its predecessor. On the other hand, if it branches right, it has at most one less vertically and at most the same horizontally as its predecessor. Furthermore, the only way it can be exactly *one* less than its predecessor is if it starts vertically from the same row and winds up on a row with the same number of dots as its predecessor wound up on. In other words, as long as the right angles are being created from the same original row of one odd part at the beginning and rows of equal length at the end, the sequence continues.

So in the example above, $14$ arises by starting on the row with one dot and ending on the row of the first thirteen. The next part starts on a row with three dots; so automatically we get a new sequence starting with $12$. The third part starts on the same row and also ends on a thirteen-dot row; so we get $11$, that is, no new sequence. This analysis reveals that the $8$, $7$, and $6$ each arise by starting on the row with nine dots and winding up on rows of thirteen dots. The $4$ and $3$ run from the nine-dot row to the eleven-dot row; hence, a new sequence. Finally the $1$ starts and ends on the nine-dot row. The inverse mapping appears at first to be somewhat tricky, but it is fairly easily constructed.

This analysis does, in fact, establish the proof of Sylvester's refinement of Euler's theorem.

---

**EXERCISE**

110. Prove that the number of partitions of $n$ into distinct parts with largest part $k$ equals the number of partitions of $n$ into odd parts, where $k$ equals the number of parts plus one half of the largest part less one. (Difficulty rating: 2)

---

## 9.2 Fine's refinement

In the early 1940s, Freeman Dyson introduced the *rank* of a partition, the largest part minus the number of parts. Hence, the rank of $4 + 3 + 2 + 2 + 1 + 1$ is $4 - 6 = -2$. Dyson's motivation in finding the rank lies in his quest for a combinatorial explanation of Ramanujan's theorem that $p(5n + 4)$ is always divisible by 5. Dyson wanted to find some parameter associated with each partition that would naturally split the partitions of $5n + 4$ into five equinumerous classes. He conjectured (and Atkin and Swinnerton-Dyer proved) that classifying partitions according to their rank modulo 5 achieved his objective.

A few years later, Nathan Fine observed that it is possible to refine Euler's theorem using this concept.

**Theorem 13 (Fine's refinement)** *The number of partitions of $n$ into distinct parts with rank either $2r$ or $2r + 1$ equals the number of partitions of $n$ into odd parts with largest part $2r + 1$.*

Thus we may group the twelve partitions of 11 into distinct parts and odd parts into six subsets as follows:

$$
\begin{array}{cc}
11 & 11 \\
10 + 1 & 9 + 1 + 1 \\
9 + 2, 8 + 3 & 7 + 3 + 1, 7 + 1 + 1 + 1 \\
8 + 2 + 1, 7 + 4, 7 + 3 + 1, 6 + 5 & 5 + 5 + 1, 5 + 3 + 3, 5 + 3 + 1 + 1 + 1, 5 + 1 + \cdots + 1 \\
6 + 4 + 1, 6 + 3 + 2, 5 + 4 + 2, & 3 + 3 + 3 + 1 + 1, 3 + 3 + 1 + \cdots + 1, 3 + 1 + 1 \cdots + 1 \\
5 + 3 + 2 + 1 & 1 + 1 + \cdots + 1
\end{array}
$$

Andrews (1983) found a very simple algorithm that provides a bijection between partitions of $n$ into odd parts with largest part $2r + 1$ and partitions of $n$ into distinct parts and rank either $2r$ or $2r + 1$. The map is as follows:

Let $\pi$ be a partition with distinct parts and rank either $2r$ or $2r + 1$. Delete the largest part of $\pi$ and add 1 to each of the remaining parts if the original rank was $2r$. If the original rank was $2r + 1$, then perform the same deletion of the largest part, add 1 to each remaining part and insert 1 as a part. Note that in each case, the number being partitioned has been reduced by $2r + 1$. Furthermore, the resulting new partition has rank $\leq 2r + 1$.

Now repeat this transformation until the partition is completely emptied. As a result, you will have produced a nonincreasing sequence of odd numbers, each $\leq 2r + 1$, which partition $n$ as desired. Furthermore, the mapping is clearly reversible, which means the bijection proves Fine's refinement.

To see the bijection in operation, we shall treat the case $n = 11$, $r = 2$ (so $2r + 1 = 5$). There are four partitions in each class, and the transformations are as follows (where an $x$ signifies a 1 added to a part):

$$
\begin{array}{c}
\cdots\cdots \\
\cdot\cdot \\
\cdot
\end{array}
\xrightarrow{\phantom{x}}
\begin{array}{c}
x\cdot\cdot \\
x\cdot \\
x \\
5
\end{array}
\xrightarrow{\phantom{x}}
\begin{array}{c}
x\cdot\cdot \\
x\cdot \\
1
\end{array}
\xrightarrow{\phantom{x}}
\begin{array}{c}
x\cdot\cdot \\
x \\
1
\end{array}
$$

$$
\xrightarrow{\phantom{x}}
\begin{array}{c}
x\cdot \\
1\ \ x
\end{array}
\xrightarrow{\phantom{x}}
\begin{array}{c}
x\cdot \\
1
\end{array}
\xrightarrow{\phantom{x}}
\begin{array}{c}
x \\
1
\end{array}
\xrightarrow{\phantom{x}} \phi
$$

so $8 + 2 + 1 \longrightarrow 5 + 1 + 1 + 1 + 1 + 1 + 1$,

$$\cdots\cdots\cdots \longrightarrow x \cdots \; \longrightarrow x \cdot \longrightarrow x \cdot \longrightarrow x \longrightarrow \phi$$
$$\cdots\cdots \quad 5 \quad x \qquad\qquad 3 \quad x \quad 1 \qquad 1 \qquad 1$$

so $7 + 4 \longrightarrow 5 + 3 + 1 + 1 + 1$,

$$\cdots\cdots\cdots$$
$$\cdots \qquad\qquad \longrightarrow x \cdots \longrightarrow x \cdots \longrightarrow \phi$$
$$\cdot \qquad\qquad\quad 5 \quad x \cdot \quad 3 \qquad 3$$

so $7 + 3 + 1 \longrightarrow 5 + 3 + 3$,

$$\cdots\cdots\cdots \longrightarrow x \cdots\cdots \longrightarrow x \longrightarrow \phi$$
$$\cdots\cdots \quad 5 \qquad\qquad\quad 5 \qquad 1$$

so $6 + 5 \longrightarrow 5 + 5 + 1$.

## 9.3 Lecture hall partitions

We shall now present a refinement of Euler's identity that is of quite recent origin. In 1995, Eriksson was working on combinatorial representations of some algebraic entities called *Coxeter groups*. As an utterly unexpected byproduct of this work, he stumbled upon a partition identity that later came to be known as the *lecture hall partition theorem*. The idea of the name is the architectural restrictions on lecture halls: Fix a number $N$ of rows of seats in a lecture hall. From every seat, there shall be a clear view of the speaker, without obstruction from the seats in front:

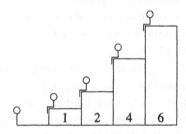

The sequence of seat heights of this lecture hall corresponds to the lecture hall partition $(1, 2, 4, 6)$. (The name *lecture hall* is also reminiscent of some famous combinatorialists, Marshall Hall and Philip Hall, whose names have been connected to many important theorems.) Formally, the set of lecture hall partitions of length $N \geq 1$ is

$$\mathcal{L}_N = \left\{ \lambda_1 + \cdots + \lambda_N : 0 \leq \frac{\lambda_1}{1} \leq \frac{\lambda_2}{2} \leq \cdots \leq \frac{\lambda_N}{N} \right\}.$$

Observe that some of the $N$ parts may be empty; if we just disregard the empty parts, we have an integer partition with the parts in increasing order. For example, $1 + 3 + 6$ is a lecture hall partition of length 3, since

$$\frac{1}{1} \leq \frac{3}{2} \leq \frac{6}{3}.$$

We are allowed to have empty parts, so $1 + 3 + 6$ could also be a lecture hall partition of length 4 – and it is:

$$\frac{1}{2} \leq \frac{3}{3} \leq \frac{6}{4}.$$

On the other hand, $1 + 4 + 5$ is not a lecture hall partition of length 3, for

$$\frac{4}{2} \not\leq \frac{5}{3}.$$

The complete set of lecture hall partitions of $n = 10$ of length $N = 3$ consists of $10, 1 + 9, 2 + 8, 3 + 7, 4 + 6, 1 + 2 + 7$, and $1 + 3 + 6$. To obtain the set of lecture hall partitions of $n = 10$ of length $N = 4$, we just add $1 + 2 + 3 + 4$ and $2 + 3 + 5$ to the previous list.

---

**EXERCISES**

111. List all lecture hall partitions of $n = 12$ for lengths $N = 1, 2, 3, 4$. (Difficulty rating: 1)

112. Explain why $\mathcal{L}_N$ is always a subset of $\mathcal{L}_{N+1}$. (Difficulty rating: 2)

113. Prove that lecture hall partitions always have distinct (non-zero) parts. (Difficulty rating: 1)

---

The identity for lecture hall partitions was proved by Bousquet-Mélou and Eriksson (1997a, 1997b, 1999) and goes as follows:

**Theorem 14 (lecture hall partition theorem)** *For a fixed length $N$ of the lecture hall, the number of lecture hall partitions equals the number of partitions into odd parts smaller than $2N$. In other words,*

$$p(n \mid lecture\ hall\ of\ length\ N) = p(n \mid odd\ parts < 2N).$$

For example, for length $N = 3$, we saw that the number of lecture hall partitions of $n = 10$ is seven. As expected, there are seven partitions of 10 into parts in $\{1, 3, 5\}$, namely, $5 + 5, 5 + 3 + 1 + 1, 5 + 1 + 1 + 1 + 1 + 1, 3 + 3 + 3 + 1,$ $3 + 3 + 1 + 1 + 1 + 1, 3 + 1 + 1 + 1 + 1 + 1 + 1 + 1$, and $1 + 1 + 1 + 1 + 1 + 1 + 1 + 1 + 1 + 1$. Just as the Rogers-Ramanujan identity, the lecture hall partition theorem seems to be a deep result in the sense that no really easy proof exists.

---

**EXERCISES**

114. By this stage, you have seen scores of partition identities. In your opinion, what is the most interesting feature of the lecture hall partition theorem when compared with previous identities? (Difficulty rating: 1)

115. Prove the lecture hall partition theorem bijectively for $N = 1$. (Difficulty rating: 1)

116. Prove the lecture hall partition theorem bijectively for $N = 2$. (Difficulty rating: 2)

117. Try to prove the lecture hall partition theorem bijectively for $N = 3$. (Difficulty rating: 3)

---

Now, in what sense is this result a refinement of Euler's identity? Well, Euler says that the number of partitions of $n$ into distinct parts equals the number of partitions of $n$ into odd parts. The lecture hall partition theorem is a "finite version" of this result: The set of partitions into odd parts smaller than $2N$ is equinumerous with the set of partitions having at most $N$ distinct parts and satisfying the additional condition of lecture hall–ness.

Euler's identity is the limiting case of the lecture hall partition theorem as $N$ tends to infinity: For a fixed $n$, if we choose $N$ large enough, then the partitions into odd parts smaller than $2N$ will in fact be all possible partitions into odd parts. On the other hand, the lecture hall partitions of $n$ for large $N$ must satisfy conditions of the type

$$\frac{\lambda_{N-k}}{N-k} \leq \frac{\lambda_{N-k+1}}{N-k+1},$$

where the numerators are much smaller than the denominators for non-zero parts, so that it is sufficient (and, of course, necessary) that $\lambda_{N-k} < \lambda_{N-k+1}$ for the inequality to hold. In other words, the lecture hall partitions of $n$ of length $N$ for large $N$ are all partitions of $n$ into distinct parts. Hence, Euler's identity follows from the lecture hall partition theorem when $N$ tends to infinity.

---

**EXERCISES**

118. If $\lambda$ is the partition $\lambda_1 + \lambda_2 + \cdots + \lambda_N$ with parts written in increasing order, define a new partition statistic $s(\lambda)$ to be the alternating sum

$\lambda_N - \lambda_{N-1} + \lambda_{N-2} - \lambda_{N-3} + \dots$ For example, if $\lambda$ is the partition $1 + 3 + 6 + 6 + 7$, then $s(\lambda) = 7 - 6 + 6 - 3 + 1$. Explain why $s(\lambda)$ is always a nonnegative number and is always positive if $\lambda$ is a partition into distinct parts. (Difficulty rating: 1)

119. There is a refinement of the lecture hall partition theorem saying that for fixed positive integers $N$ and $S$, the number of lecture hall partitions $\lambda$ of length $N$ with $s(\lambda) = S$ equals the number of partitions into $S$ odd parts smaller than $2N$. Verify this statement for partitions of 19 with $N = 5$ and $S = 3$ by listing the two sets of partitions. (Difficulty rating: 1)

120. As $N$ tends to infinity, what refinement of Euler's identity is derived from the result in the previous exercise? (Difficulty rating: 1)

121. (Kim and Yee, 1999) Try to find a direct proof of the assertion in Exercise 120 by adapting the Sylvester construction from the beginning of this chapter. (Difficulty rating: 3)

---

We shall now take the first step in a proof of the lecture hall partition theorem, leading to a method to prove the theorem for any given $N$. Let us study the partition $2 + 5 + 11$, which is a lecture hall partition of length 3, since

$$\frac{2}{1} \le \frac{5}{2} \le \frac{11}{3}.$$

Observe that if we subtract 1 from the first part, 2 from the second, and 3 from the third, then what remains is still a lecture hall partition: $1 + 3 + 8$. Applying the same trick again yields $0 + 1 + 5$. Now we cannot subtract 1 from the first part anymore, nor 2 from the second part. But we can subtract 3 from the third part to obtain $0 + 1 + 2$. No further subtractions of 3 from the third part are possible. We say that $0 + 1 + 2$ is a *reduced* lecture hall partition of length 3.

In general, a lecture hall partition is reduced if the lecture hall property is destroyed whenever $k$ is subtracted from the $k$th part, for all $k = 1, 2, \dots, N$. There exist six reduced lecture hall partitions of length 3:

$$0 + 0 + 0, \ 0 + 0 + 1, \ 0 + 0 + 2, \ 0 + 1 + 2, \ 0 + 1 + 3, \ 0 + 1 + 4.$$

Every nonreduced lecture hall partition of length 3 can be reduced by a sequence of subtractions of the blocks $1 + 2 + 3, 0 + 2 + 3$, and $0 + 0 + 3$. Conversely, starting from any reduced lecture hall partition, any sequence of additions of blocks $1 + 2 + 3, 0 + 2 + 3$, and $0 + 0 + 3$ will result in a unique lecture hall partition. (Why?) The weights of the reduced lecture hall partitions (of which exactly one must be chosen) are, respectively, 0, 1, 2, 3, 4, and 5. The weights of the repeated blocks are 6, 5, and 3. Therefore, the generating function for lecture hall partitions of length 3 can be expressed and manipulated as follows:

$$\frac{q^0 + q^1 + q^2 + q^3 + q^4 + q^5}{(1-q^3)(1-q^5)(1-q^6)} = \frac{(1-q^6)/(1-q)}{(1-q^3)(1-q^5)(1-q^6)}$$

$$= \frac{1}{(1-q)(1-q^3)(1-q^5)},$$

which is the generating function for partitions into odd parts smaller than 6, as asserted by the lecture hall partition theorem.

---

**EXERCISES**

122. List the two reduced lecture hall partitions of length 2. (Difficulty rating: 1)

123. List the twenty-four reduced lecture hall partitions of length 4. (Difficulty rating: 1)

124. Explain why there are always $N!$ reduced lecture hall partitions of length $N$. (Difficulty rating: 1)

125. Explain why partwise addition of blocks $1 + 2 + \cdots + N, 2 + 3 + \cdots + N, 3 + 4 + \cdots + N$, and so on, to a lecture hall partition always produces a new lecture hall partition. Explain why every lecture hall partition can be constructed in this way from a reduced lecture hall partition. (Difficulty rating: 2)

126. Prove the lecture hall partition theorem for $N = 2$ and $N = 4$ using generating functions. (Difficulty rating: 2)

127. Describe what kind of result would be needed as the next step in order to prove the lecture hall partition theorem for general $N$ using reduced partitions. (Difficulty rating: 3)

128. Prove the partition identity

$p(n \mid \text{odd parts} < 2N) = p(n \mid \text{parts} \leq 2N, \text{where parts} \leq N \text{ are distinct}),$

both by generating functions and by finding a bijection. For example, for $N = 3$ and $n = 10$, we know that the left-hand number is seven. The seven partitions counted to the right are: $6 + 4, 6 + 3 + 1, 5 + 5, 5 + 4 + 1, 5 + 3 + 2, 4 + 4 + 2$, and $4 + 3 + 2 + 1$. (See Yee (2002).) (Difficulty rating: 3)

It seems to be difficult to find a direct bijection proving the lecture hall partition theorem. A rather involved bijection between the right-hand partitions of the last exercise above and lecture hall partitions was found by Bousquet-Mélou and Eriksson (1999):

Let $\mu$ be a partition into parts $\leq 2N$ with distinct parts $\leq N$. We will describe an algorithm that takes the parts of $\mu$ in increasing order and constructs step by step a lecture hall partition $\lambda$ of length $N$. Rather, what is constructed is an $N$-by-$N$ triangle $T$ of integers, the column sums of which are the parts of the lecture hall partition $\lambda$:

$$
\begin{array}{cccc}
t_{11} & t_{12} & \cdots & t_{1N} \\
 & t_{22} & \cdots & t_{2N} \\
 & & \ddots & \vdots \\
 & & & t_{NN}
\end{array}
$$

where $\lambda_i = t_{1i} + t_{2i} + \cdots + t_{ii}$. Let $\mu = \mu_1 + \cdots + \mu_r$, where the parts come in increasing order.

**Step 0.** Take all parts in $\mu$ that are less than or equal to $N$, say $\mu_1, \ldots, \mu_k$. Construct a sequence of $N$ columns of 0s and 1s forming a triangle $T^{(0)}$, where column $N - i$ has $\mu_{k-i}$ 1s, for $i = 0, 1, \ldots, k - 1$. The 1s are placed at the bottom of the columns; all other entries are 0s.

**Step $s = 1, 2, \ldots, r - k$.** From the previous triangle $T^{(s-1)}$ and the next part $\mu_{k+s}$, we construct the next triangle $T^{(s)}$ as follows. We know that $\mu_{k+s} = n + i$ for some $i = 1, \ldots, N$. Remove the $i$th row and column from $T^{(s-1)}$, yielding an $N - 1$-by-$N - 1$ triangle. Add a new $N$th column by setting

$$
t^{(s)}_{NN} = t^{(s-1)}_{ii} + 2, \qquad t^{(s)}_{jN} = t^{(s-1)}_{ji} + 2 \quad \text{for } j = 1, \ldots, i - 1
$$

and

$$
t^{(s)}_{jN} = t^{(s-1)}_{i,j+1} + 1 \quad \text{for } j = i, \ldots, n - 1.
$$

Pictorially:

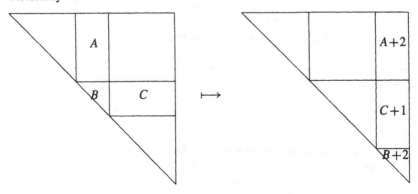

For example, let $N = 4$ and $\mu = 1 + 2 + 4 + 5 + 6 + 6$. The parts $\leq N$ are 1, 2, and 4, so in Step 0 we obtain the triangle

$$
\begin{array}{cccc}
\mathbf{0} & \mathbf{0} & \mathbf{0} & \mathbf{1} \\
1 & 1 & 1 & \\
1 & 1 & & \\
1 & & &
\end{array}
$$

Since the next part in $\mu$ is $\mu_3 = 5 = 4 + 1$, we have boldfaced row 1 and column 1 to signal that they will be removed and replaced as a new fourth column $(0 + 1, 0 + 1, 1 + 1, 0 + 2) = (1, 1, 2, 2)$ in Step 1:

$$
\begin{array}{cccc}
1 & \mathbf{1} & \mathbf{1} & \mathbf{1} \\
\mathbf{1} & \mathbf{1} & \mathbf{1} & \\
1 & 2 & & \\
2 & & &
\end{array}
$$

In Step 2, we use $\mu_4 = 6 = 4 + 2$. Hence, we now do the maneuver with row 2 and column 2, yielding a new last column $(1 + 2, 1 + 1, 1 + 1, 1 + 2) = (3, 2, 2, 3)$:

$$
\begin{array}{cccc}
1 & \mathbf{1} & 1 & 3 \\
\mathbf{1} & \mathbf{2} & \mathbf{2} & \\
2 & 2 & & \\
3 & & &
\end{array}
$$

Finally, in Step 3, we use $\mu_5 = 6 = 4 + 2$, so once again, we work with row 2 and column 2, replacing them by a new last column $(1 + 2, 2 + 1, 2 + 1, 1 + 2) = (3, 3, 3, 3)$:

$$
\begin{array}{cccc}
1 & 1 & 3 & 3 \\
2 & 2 & 3 & \\
3 & 3 & & \\
3 & & &
\end{array}
$$

We ended up with the lecture hall partition $\lambda = 1 + 3 + 8 + 12$ of length $N = 4$. The above procedure is indeed a bijection! However, the proof of this involves Coxeter groups and is not suitable for this book.

We note that there is an alternative bijection given by Yee (2001).

---

## EXERCISES

129. Run the bijection for all partitions with $n = 10$ and $N = 3$. (Difficulty rating: 1)

130. What does the bijection specialize to for $N = 2$? Prove that it works in this special case! (Difficulty rating: 3)

# Chapter 10

## Plane partitions

Up to this point, you have been working with partitions, such as $5 + 4 + 4 + 2$, wherein the summands are written in a line. We may reasonably call these *one-dimensional partitions*, suggesting that there are higher dimensional partitions. In this chapter, we shall study two-dimensional partitions, or *plane partitions*.

Now instead of a row of summands, we consider an array of rows of integers that is left justified, wherein there is non-increase along rows and columns. For example, the thirteen plane partitions of 4 are:

$$
\begin{array}{llllllll}
4, & 3 & 1, & 3, & 2 & 2, & 2, & 2 & 1 & 1, & 2 & 1, & 2, \\
   &   & 1  &    & 1 &    & 2  &   &   & 1  &   & 1  & 1 \\
   &   &    &    &   &    &    &   &   &    &   &    & 1
\end{array}
$$

$$
\begin{array}{lllllllll}
1 & 1 & 1 & 1, & 1 & 1 & 1, & 1 & 1, & 1 & 1, & 1 \\
  &   &   & 1  &   &   & 1  & 1 & 1  &   & 1  \\
  &   &   &    &   &   &    & 1 &    &   & 1  \\
  &   &   &    &   &   &    &   &    &   & 1
\end{array}
$$

---

### Highlights of this chapter

- Plane partitions can be represented by three-dimensional Ferrers boards, which in turn can be viewed as rhombus tilings of a hexagon.
- MacMahon found beautiful formulas for the generating functions of plane partitions and restricted plane partitions.

---

## 10.1 Ferrers graphs and rhombus tilings

As you recall from Chapter 3, one-dimensional partitions have a geometric representation called a *Ferrers board*. For example, the representation of

$4 + 4 + 2 + 1 + 1$ is

We see the use of a two-dimensional geometric figure to represent our one-dimensional partitions, so it is not surprising that plane partitions have three-dimensional Ferrers boards. Now instead of squares juxtaposed linearly, we stack cubes over the position of each part. Thus the plane partition

$$
\begin{array}{cccc}
4 & 3 & 2 & 1 \\
3 & 1 & 1 \\
2 \\
2
\end{array}
$$

has a Ferrers 3D-board looking like this (generated by Excel):

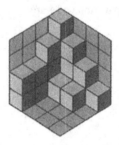

This mode of representation leads to the marvellous observation that the above array actually defines the tiling (i.e., nonoverlapping covering) of a regular hexagon, with sides of length 4, by unit rhombi (with interior angles 60°, 120°, 60°, and 120°, in succession). Namely,

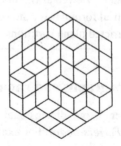

Indeed, you can see that if you shade each rhombus according to its orientation, then the picture that emerges from the tiled hexagon of side $= n$ is that of the Ferrers 3D-board of a plane partition wherein the number of rows, the number of columns, and the largest part are each $\leq n$.

Adding an extra cube to the 3D-board corresponds to rotating the tiling of a unit hexagon:

This "new view" of plane partitions has grown into an independent subject, the theory of rhombus tilings, which now has an extensive literature and has produced numerous new techniques and unsolved problems of its own.

---

**EXERCISES**

131. Draw the twenty tilings of the regular hexagon of side 2 by unit rhombi. (Difficulty rating: 1)

132. Prove that the bijection between plane partitions and rhombus tilings of the hexagon (as described in this section) actually holds. Hint: It is clear that each Ferrers 3D-board of a plane partition produces such a tiling. Can such a tiling arise that does not correspond to the Ferrers 3D-board of a plane partition? (Difficulty rating: 2)

---

## 10.2 MacMahon's formulas

Plane partitions have generating functions associated with them of wonderful simplicity. Suppose we let $pp(n)$ denote the total number of plane partitions of $n$. MacMahon proved the following beautiful formula:

**Theorem 15 (MacMahon's plane partition formula)** *The generating function for the number of plane partitions is*

$$\sum_{n=0}^{\infty} pp(n)q^n = \prod_{n=1}^{\infty} \frac{1}{(1-q^n)^n}$$

*for* $|q| < 1$.

Major P. A. MacMahon began his mathematical work in invariant theory, one of the hot topics of nineteenth-century mathematicians. This subject in turn led him to the theory of partitions because of the auxiliary role played by partitions in invariant theory. As he attempted to elucidate the intricacies of compositions (ordered partitions) and multipartite partitions (partitions of $n$-tuples of integers), his studies led him to two-dimensional or plane partitions following Sylvester's lead involving Ferrers graphs. To MacMahon's surprise (MacMahon, 1897) and delight, proving Theorem 15 – which is as elegant as Euler's beautiful generating function theorem (5.7) – turned out to be a challenge that would take him twenty years (MacMahon, 1916)!

A refinement of MacMahon's plane partition formula is obtained if we restrict ourselves to $pp_k(n)$, the number of plane partitions with at most $k$ rows. Then the generating function is given by

$$\sum_{n=0}^{\infty} pp_k(n)q^n = \prod_{n=1}^{\infty} \frac{1}{(1-q^n)^{\min(k,n)}}. \tag{10.1}$$

Note that this last result includes the case $k = 1$, that is, one-dimensional partitions.

What is surprising is the fact that these results on plane partitions, although so simple to state, are quite difficult to prove when compared to the corresponding theorems in Chapter 5 for one-dimensional partitions. In fact, we shall here be content to prove (10.1) only in the case $k = 2$. To this end, we define $\pi_r(h, j; q)$ to be the generating function for plane partitions with at most 2 rows, and satisfying the further restrictions of having at most $r$ parts in each row, with $h$ the first part on the top row and $j$ the first part on the bottom row.

For example,

$$\pi_3(2, 2; q) = q^4 + q^5 + 3q^6 + 3q^7 + 4q^8 + 3q^9 + 3q^{10} + q^{11} + q^{12}; \tag{10.2}$$

the coefficient 4 for $q^8$ enumerates the plane partitions

| 22 | 222 | 221 | 211 |
| 22 | 2   | 21  | 211 |

In fact, you can work out each $\pi_r(h, j; q)$ from the trivial formula for $r = 1$,

$$\pi_1(h, j; q) = q^{h+j}, \tag{10.3}$$

and the recursion,

$$\pi_r(h, j; q) = q^{h+j} \sum_{m=0}^{j} \sum_{n=m}^{h} \pi_{r-1}(n, m; q). \tag{10.4}$$

For example, the step from $r = 1$ to $r = 2$ looks as follows:

$$\pi_2(h, j; q) = q^{h+j} \sum_{m=0}^{j} \sum_{n=m}^{h} q^{n+m}$$

$$= q^{h-j} \sum_{m=0}^{j} q^m \cdot \frac{q^m - q^{h+1}}{1 - q}$$

$$= \frac{q^{h+j}}{(1-q)} \left( \frac{1 - q^{2j+2}}{1 - q^2} - \frac{q^{h+1}(1 - q^{j+1})}{(1-q)} \right)$$

$$= \frac{q^{h+j}(1 - q^{j+1})}{(1-q)(1-q^2)} \left( (1 + q^{j+1}) - q^{h+1}(1+q) \right)$$

$$= \frac{q^{h+j}(1 - q^{j+1})}{(1-q)(1-q^2)} \left( (1 - q^{h+1})(1+q) - q(1 - q^j) \right)$$

$$= q^{h+j} \left( \begin{bmatrix} h+1 \\ 1 \end{bmatrix} \begin{bmatrix} j+1 \\ 1 \end{bmatrix} - q \begin{bmatrix} j+1 \\ 2 \end{bmatrix} \right).$$

---

## EXERCISES

133. Show that

$$\pi_3(h, j; q) = q^{h+j} \left( \begin{bmatrix} h+2 \\ 2 \end{bmatrix} \begin{bmatrix} j+2 \\ 2 \end{bmatrix} - q \begin{bmatrix} h+2 \\ 1 \end{bmatrix} \begin{bmatrix} j+2 \\ 3 \end{bmatrix} \right).$$

(Difficulty rating: 2)

134. Show that the result in the previous exercise is consistent with the recursion (10.4). (Difficulty rating: 2)

---

## 10.3 The formula for $\pi_r(h, j; q)$

Following the approach of L. Carlitz (1967), we are now ready to prove by mathematical induction on $r$ that

$$\pi_r(h, j; q) = q^{h+j} \left( \begin{bmatrix} h+r-1 \\ r-1 \end{bmatrix} \begin{bmatrix} j+r-1 \\ r-1 \end{bmatrix} \right.$$

$$\left. - q \begin{bmatrix} h+r-1 \\ r-2 \end{bmatrix} \begin{bmatrix} j+r-1 \\ r \end{bmatrix} \right). \tag{10.5}$$

We observe that the cases $r = 1$ and $r = 2$ were proved in the previous section. In order to handle the induction step, we shall require the following formula (which was given as an exercise in Chapter 7):

$$\sum_{j=0}^{N} q^j \begin{bmatrix} M + j \\ j \end{bmatrix} = \begin{bmatrix} N + M + 1 \\ M + 1 \end{bmatrix}. \tag{10.6}$$

So we assume (10.5) is true up to and including a given $r$. Then the induction step goes like this:

$$\pi_{r+1}(h, j; q)$$

$$= q^{h+j} \sum_{m=0}^{j} \sum_{n=m}^{h} \pi_r(n, m; q) \quad \text{(by (10.4))}$$

$$= q^{h+j} \sum_{m=0}^{j} \sum_{n=m}^{h} q^{n+m} \left( \begin{bmatrix} n + r - 1 \\ r - 1 \end{bmatrix} \begin{bmatrix} m + r - 1 \\ r - 1 \end{bmatrix} \right.$$

$$\left. - q \begin{bmatrix} n + r - 1 \\ r - 2 \end{bmatrix} \begin{bmatrix} m + r - 1 \\ r \end{bmatrix} \right)$$

$$= q^{h+j} \sum_{m=0}^{j} \left( \sum_{n=m}^{h} q^{n+m} \begin{bmatrix} n + r - 1 \\ r - 1 \end{bmatrix} \begin{bmatrix} m + r - 1 \\ r - 1 \end{bmatrix} \right.$$

$$\left. - \sum_{n=m+1}^{h+1} q^{n-m} \begin{bmatrix} m + r - 1 \\ r \end{bmatrix} \begin{bmatrix} n + r - 2 \\ r - 2 \end{bmatrix} \right)$$

$$= q^{h+j} \left( \sum_{m=0}^{j} q^m \begin{bmatrix} m + r - 1 \\ r - 1 \end{bmatrix} \left( \begin{bmatrix} r + h \\ r \end{bmatrix} - \begin{bmatrix} r + m - 1 \\ r \end{bmatrix} \right) \right.$$

$$\left. - \sum_{m=0}^{j} q^m \begin{bmatrix} m + r - 1 \\ r \end{bmatrix} \left( \begin{bmatrix} r + h \\ r - 1 \end{bmatrix} - \begin{bmatrix} r + m - 1 \\ r - 1 \end{bmatrix} \right) \right) \quad \text{(by (10.6))}$$

$$= q^{h+j} \left( \begin{bmatrix} r + h \\ r \end{bmatrix} \sum_{m=0}^{j} q^m \begin{bmatrix} m + r - 1 \\ r - 1 \end{bmatrix} \right.$$

$$\left. - \begin{bmatrix} r + h \\ r - 1 \end{bmatrix} \sum_{m=0}^{j} q^m \begin{bmatrix} m + r - 1 \\ r \end{bmatrix} \right)$$

$$= q^{h+1} \left( \begin{bmatrix} r + h \\ r - 1 \end{bmatrix} \begin{bmatrix} j + r \\ r \end{bmatrix} - \begin{bmatrix} r + h \\ r - 1 \end{bmatrix} \begin{bmatrix} r + j \\ r + 1 \end{bmatrix} \right) \quad \text{(by (10.6))}.$$

This is precisely the required formula for $\pi_{r+1}(h, j; q)$. Hence, (10.5) has been established by mathematical induction.

As a corollary, we immediately deduce MacMahon's formula for $pp_2(n)$, the number of plane partitions with at most two rows:

$$\sum_{n=0}^{\infty} pp_2(n)q^n = \lim_{\substack{r \to \infty \\ j \to \infty}} q^{-h-j}\pi_r(j, j; q)$$

$$= \prod_{n=1}^{\infty} \frac{1}{(1-q^n)^2} - q \prod_{n=1}^{\infty} \frac{1}{(1-q^n)^2}$$

$$= (1-q) \prod_{n=1}^{\infty} \frac{1}{(1-q^n)^2}$$

$$= \prod_{n=1}^{\infty} \frac{1}{(1-q^n)^{\min(2,n)}}.$$

## EXERCISES

135. Show that $\pi_2(j, j; q) = q^{2j}\begin{bmatrix} j+2 \\ 2 \end{bmatrix}$. (Difficulty rating: 2)

136. More generally show that

$$\pi_r(j, j; q) = q^{2j} \begin{bmatrix} j+r \\ r \end{bmatrix} \begin{bmatrix} j+r \\ r-1 \end{bmatrix} \Big/ \begin{bmatrix} j+r \\ 1 \end{bmatrix}.$$

(Difficulty rating: 2)

# Chapter 11

## Growing Ferrers boards

In the past decade, there has been a surge of interest in random processes on combinatorial objects. In this chapter, we deal with recent developments in random domino tilings and random Ferrers boards.

---

**Highlights of this chapter**

- The set of all integer partitions can be partially ordered according to the rule that one partition comes before another if the Ferrers board of the latter can be obtained by addition of squares to the Ferrers board of the former.
- With this dynamic view of partitions growing by stepwise addition of squares, we obtain objects called standard tableaux by keeping track of the order in which the squares of a Ferrers board have been added.
- With a simple rule for random growth of partitions, the shape of the rim of the Ferrers board tends to a quarter-circle. This growth process of partitions turns out to be closely related to random domino tilings of a shape called the Aztec diamond. The arctic circle theorem of Jockusch, Propp and Shor describes how four quarter-circles form a circular boundary ("the arctic circle") between well-ordered ("frozen") tiles in the corners and an inner chaos ("temperate zone") in the middle.

---

## 11.1 Random partitions

How, for instance, can one generate random partitions of $n$ so that all partitions are chosen with equal probability? For small $n$, say $n = 4$, it is easy. Just list all partitions:

Then we pick a random number between 1 and $p(n)$ – in this case between 1 and 5 – and choose the corresponding partition. But if $n$ is large, say $n = 1000$, there are so many partitions that it is unreasonable to list them all before we make our choice. How, then, can we choose a partition of 1000 with uniform probability?

An idea that seems sensible is to construct a random Ferrers board by adding squares stepwise at randomly chosen inner corners. An example of how such a process could run is

It turns out that this random growth process does not generate all partitions with equal probability. In fact, some partitions will be extremely improbable outcomes compared with others. So random growth of partitions fails to generate uniformly random partitions, nevertheless it is well worth studying – it gives rise to more exciting mathematics than the original question! We will conclude this chapter with a quite spectacular recent result on random tilings with dominoes, called the *arctic circle theorem*, where growing Ferrers boards play a fundamental role.

---

**EXERCISE**

137. Think about other ways to tackle the problem of generating uniformly random partitions of a given large integer. (Difficulty rating: 3)

---

## 11.2 Posets of partitions

Let us start with a short discussion of total and partial orderings. The set of nonnegative integers has a total ordering by size: $0 < 1 < 2 < 3 \ldots$. Any number can be compared with any other number, and the comparison will always reveal that one of the numbers is larger than the other one. For integer points in the plane, we can try to define an ordering in a similar way: Say that $(a, b) \preceq (m, n)$ if $a \leq m$ and $b \leq n$, that is, if coordinatewise comparison gives an unambiguous result. For example, under this ordering, we have

$(0, 0) \preceq (1, 0) \preceq (1, 1) \preceq (1, 2)$. But if we compare $(1, 0)$ to $(0, 1)$, then neither point comes before the other in this order. We say that the set of integer points in the plane is *partially ordered* by $\preceq$, since only some pairs are comparable.

There are several useful partial orderings of integer partitions. We will describe the so-called *usual order*. It is most easily defined on the Ferrers boards. One Ferrers board precedes another in the usual order if it can be contained in the other. (You can think of the upper left corners as coinciding.) For instance, the board

contains the boards

A partially ordered set is usually referred to as a *poset*. There is a convenient way of representing posets graphically, called the *Hasse diagram* of a poset. Hasse diagrams are drawn according to two simple rules. First, greater elements are placed above smaller elements on the paper. Second, we draw a line between two elements if they are comparable under the partial ordering and if there exists no element in between. The point is to draw no unnecessary lines. If two elements are comparable, then there exists a path in the diagram going upward from the smaller to the greater element. Here is the Hasse diagram of the poset under the usual order of partitions of all integers less than or equal to four:

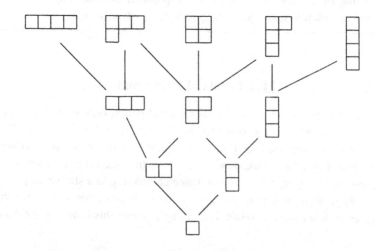

In the Hasse diagram of the usual order of integer partitions, there will be a line between boards $\lambda$ and $\mu$ if and only if we can obtain $\mu$ by filling an inner corner of $\lambda$ with a square. (See Chapter 3.) Therefore, an upward path from the bottom element to a partition $\lambda$ is a complete instruction for constructing the Ferrers board of $\lambda$ by adding one square at a time. For example, there are three paths leading to $3 + 1$:

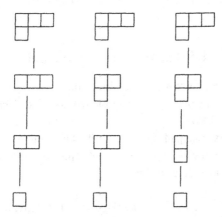

Let us denote the usual order on partitions by $\preceq$. Thus the rightmost path above implies that $1 \preceq 1 + 1 \preceq 2 + 1 \preceq 3 + 1$. Since $1 + 1 + 1$ is not less than $3 + 1$ under the usual order, we write $1 + 1 + 1 \npreceq 3 + 1$.

---

## EXERCISES

138. Extend the Hasse diagram of the usual order up to partitions of 5. (Difficulty rating: 1)

139. Characterize those partitions for which there is a unique path from the bottom element. (Difficulty rating: 2)

140. Let $\lambda'$ and $\mu'$ denote the conjugates of partitions $\lambda$ and $\mu$. Show that if $\lambda \preceq \mu$, then $\lambda' \preceq \mu'$. (Difficulty rating: 2)

141. Try to define a total ordering of integer partitions. (Hint: There is a total ordering used in dictionaries.) Draw its Hasse diagram for all partitions of integers up to 4. (Difficulty rating: 2)

142. An alternative to the usual order is the *dominance order*, which we denote by $\trianglelefteq$. Let parts be ordered in decreasing size, so that $\lambda = \lambda_1 + \lambda_2 + \ldots$, where $\lambda_1 \geq \lambda_2 \geq \ldots$ (including an unlimited supply of zero parts). Then the dominance order is defined by $\lambda \trianglelefteq \mu$ if the following inequalities all are satisfied:

$\lambda_1 \leq \mu_1, \lambda_1 + \lambda_2 \leq \mu_1 + \mu_2, \lambda_1 + \lambda_2 + \lambda_3 \leq \mu_1 + \mu_2 + \mu_3$, etc.

Draw the Hasse diagram for partitions of integers up to 5 under dominance order. (Difficulty rating: 1)

143. Is it true that if $\lambda$ precedes $\mu$ in the usual order, then it must also do so in the dominance order? Is the converse true? (Difficulty rating: 2)

144. Show that conjugation reverses the dominance order of partitions of the same size, that is, if $\lambda \trianglelefteq \mu$, then $\mu' \trianglelefteq \lambda'$. (Difficulty rating: 2)

## 11.3 The hook length formula

The aim of this chapter is to say something about what a typical Ferrers board that grows randomly by one square at a time will look like. Therefore, we need to know how many paths the growth process can follow to end up in a given Ferrers board. Every such path can be described by numbering every square in the board according to when it was included. Thus, the three paths ending the previous section can be described by

$$\begin{array}{|c|c|c|}\hline 1 & 2 & 3 \\\hline 4 \\\cline{1-1}\end{array}, \quad \begin{array}{|c|c|c|}\hline 1 & 2 & 4 \\\hline 3 \\\cline{1-1}\end{array}, \quad \text{and} \begin{array}{|c|c|c|}\hline 1 & 3 & 4 \\\hline 2 \\\cline{1-1}\end{array}, \text{ respectively.}$$

The objects that appear in this way are Ferrers boards with the squares filled with the numbers 1 up to the number of squares, in such a way that the numbers in every row and column are increasing. These number tables are called *standard tableaux* (or standard Young tableaux) and are objects of great importance in various branches of mathematics. By counting paths from the bottom and up to the $n$th level in the Hasse diagram of the usual order of partitions, we can conclude that the number of standard tableaux with $n$ squares is 1, 2, 4, and 10 for $n = 1, 2, 3,$ and 4, respectively.

**EXERCISES**

145. Find the number of standard tableaux with five squares. (Difficulty rating: 1)

146. A procedure known as *row insertion* can be used to insert a new number $N$ into a standard tableau (possibly with some gaps in the set of numbers used). Row insertion works as follows: If $N$ is greater than the rightmost number of the top row, then the top row is extended by a new square containing $N$. Otherwise the top row contains some smallest number $N'$ greater than $N$. Replace $N'$ by $N$ and iterate the procedure of inserting $N'$ into the second row. Here is an example of row insertion:

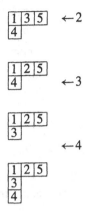

Explain why row insertion always results in a new standard tableau (possibly with some gaps in the set of numbers used). (Difficulty rating: 2)

147. Explain how row insertion can be played backwards, starting with any corner number. (Difficulty rating: 2)

148. Let $s_n$ denote the number of standard tableaux with $n$ squares. Prove that $s_n$ satisfies the recursion $s_n = s_{n-1} + (n-1)s_{n-2}$ for $n \geq 2$, and $s_1 = s_0 = 1$. Hint: The first term counts standard tableaux where the number $n$ is in a square (at the bottom) of the leftmost column. The second term counts standard tableaux where the number $n$ is not in the leftmost column. In this case there is some other number at the bottom of the column directly to the left of $n$; play row insertion backwards, starting with this number. (Difficulty rating: 3)

149. Explain why the same recursion also gives the number of partitions of the set $\{1, 2, \ldots, n\}$ into blocks of size 1 or 2. For example, for $n = 3$ there are the four relevant set partitions $\{12, 3\}$, $\{13, 2\}$, $\{1, 23\}$ and $\{1, 2, 3\}$. (Difficulty rating: 2)

---

There is a famous formula for the number of standard tableaux of a given shape $\lambda$. It is called the *hook length formula*, for reasons that will soon become obvious. Any square in a Ferrers board determines a *hook* consisting of that square, all squares to the right of it, and all squares below it. The picture below shows the hook of the marked square:

The *hook length* is the number of squares in the hook. We can record all hook lengths of a shape in the respective squares; for example:

| 4 | 2 | 1 |
|---|---|---|
| 1 |   |   |

and

| 7 | 5 | 3 | 2 |
|---|---|---|---|
| 6 | 4 | 2 | 1 |
| 3 | 1 |   |   |
| 1 |   |   |   |

Let $f^\lambda$ denote the number of standard tableaux of shape $\lambda$. From our previous investigations, we know that $f^{3+1} = 3$. The hook length formula of Frame, Robinson and Thrall says that

$$f^\lambda = \frac{|\lambda|!}{\prod_{ij \in \lambda} h_{ij}^\lambda}, \tag{11.1}$$

where $h_{ij}^\lambda$ is the length of the hook for the square on row $i$ and column $j$ in shape $\lambda$. The product is taken over all squares in the Ferrers board. Thus, for $\lambda = 3 + 1$, we obtain

$$f^{3+1} = \frac{4!}{4 \cdot 2 \cdot 1 \cdot 1} = \frac{24}{8} = 3,$$

as we already knew.

There are many proofs of the hook length formula, although none really well suited for this book. A heuristic way to 'see' the formula is by considering all $|\lambda|!$ numberings of the squares, and throwing away a numbering if there is some square $ij$ which does not contain the smallest number of its hook. For every square $ij$, the probability of its number being smallest in its hook is $1/h_{ij}^\lambda$. Thus, if all these probabilities were independent, then the hook length formula follows. Sagan (1991) gives a rigorous proof along these lines.

Given this remarkable formula, we can readily deduce explicit expressions for interesting special cases. For example, the hook lengths for a staircase are:

| $2\ell$-1 | $\ddots$ | 5 | 3 | 1 |
|-----------|----------|---|---|---|
| $\ddots$  | 5        | 3 | 1 |   |
| 5         | 3        | 1 |   |   |
| 3         | 1        |   |   |   |
| 1         |          |   |   |   |

The hook length formula then immediately tells us that the number of standard tableaux of this staircase shape is

$$f^{\ell + (\ell-1) + \cdots + 1} = \frac{(\ell(\ell+1)/2)!}{1^\ell \cdot 3^{\ell-1} \cdot 5^{\ell-2} \cdots (2\ell-1)^1}.$$

## EXERCISES

150. Use the hook length formula to compute the number of standard tableaux for all shapes with six squares. (Difficulty rating: 1)

151. For every $n$ from 1 up to 6, which shape of $n$ squares can be grown in the greatest number of ways? (Difficulty rating: 1)

152. Compute $\sum_{\lambda \vdash n}(f^\lambda)^2$ for $n = 1, 2, 3, 4, 5$, and come up with a conjecture. ($\lambda \vdash n$ means that $\lambda$ is a partition of $n$.) (Difficulty rating: 2)

153. Prove the hook length formula for shapes of at most two rows. You can do it by induction. There are two cases to consider: Either the shape has one outer corner, such as ⊞⊞, in which case the previous shape in the path must have been ⊞⊞. Otherwise, the shape has two outer corners, such as ⊞⊞, in which case the number of paths is the sum of the numbers of paths to the two possible predecessors, ⊞⊞ and ⊞⊞. You must show that the hook length formula satisfies the same recursion. (Difficulty rating: 2)

154. Use the hook length formula to show that the number of standard tableaux of shape two rows of equal length $m$ is the Catalan number $C_m$ (see Chapter 3). (Difficulty rating: 2)

155. Find a bijection between the standard tableaux in the previous exercise and Ferrers boards fitting inside a staircase of nonreduced height $m$ (see Chapter 3). (Difficulty rating: 3)

## 11.4 Randomly growing Ferrers boards

Let us now test the procedure for generating a random Ferrers board that we mentioned in the introduction: In every step, choose an inner corner at random and fill it with a new square. (We refer to this random process as *type I*.) As the number $n$ of squares grows, the probability of any particular board appearing tends to zero. But is there a typical overall shape? Here are the results of three different runs of this random growth process for $n = 150$.

Although the above three samples are very different in their details, they share the property that their contours are slightly concave. Actually, Rost (1981)

proved that as $n$ grows, the contour tends to a parabolic shape (in a probabilistic sense that will be made precise later in this chapter).

Another way (which we refer to as *type II*) of randomizing the growth process is to flip a coin for every inner corner at each step. We then fill all those inner corners for which the coin came up tails. This leads to random paths in a graph consisting of the Hasse diagram of the usual order with some new edges included, corresponding to the possibilities of adding more than one square at a time:

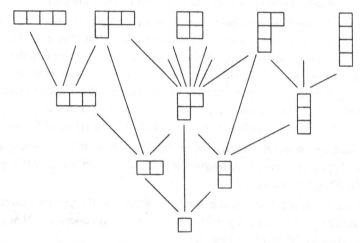

Three runs of this type II random growth process until (at least) 150 squares were added gave the following results:

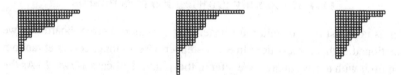

These samples may look similar to the ones from the type I process, but one can imagine that those of type II are perhaps generally a little fatter. Jockush, Propp, and Shor proved in 1995 that type II contours tend to become roughly shaped as *quarter-circles*. We will meet their result later in this chapter as the *Arctic Circle theorem*. A related study was made by Pittel (1997).

---

**EXERCISES**

156. Produce a heuristic explanation for why it is reasonable that the typical shape of a randomly grown Ferrers board (type I or type II) is concave and not convex. (Difficulty rating: 2)

157. Although Ferrers boards are two-dimensional, it is possible to model their growth one-dimensionally as follows: Start by a sequence of sufficiently many 0s followed by sufficiently many 1s. In every step, choose some occurrence of the pattern 01 in the sequence, and replace it by 10. (This is the setting Rost worked with.) Explain the connection to growing Ferrers boards! (Difficulty rating: 2)

158. Write a computer program for random growth of Ferrers boards. (It's easy!) Try playing around with the randomization. For example, does it seem to matter if the coin in the type II process is biased? Or, in the type I process, what happens if the probability of an inner corner being chosen is less if it belongs to a longer row? (Difficulty rating: 3)

## 11.5 Domino tilings

In order to proceed from the random Ferrers boards of the previous pages to the arctic circle theorem in the next section, we need an interlude about domino tilings. When you solve a jigsaw puzzle, you arrange pieces of the puzzle that fit together into a complete picture of, say, a rectangular shape. If we disregard the decorative picture, we arrive at the mathematical notion of a *tiling* of a region, in this case, a rectangle. Thus, a tiling is a covering of a region by nonoverlapping pieces (called *tiles*).

Particularly well studied are tilings by dominoes (one-by-two rectangles). Obviously, there are two domino tilings of a two-by-two square, since the two dominoes must lie either horizontally ⊟ or vertically ⊞. For a chessboard (eight-by-eight), there are 12,988,816 domino tilings! Here is one possible domino tiling of the chessboard:

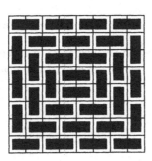

Domino tilings play a role in statistical physics (where the dominoes are called *dimers*, meaning diatomic molecules). In the early sixties, physicists Kasteleyn, Fisher and Temperley found the number of domino tilings of rectangular chessboards of arbitrary even size, say $2m$ by $2n$, to be

$$N(n, m) = \prod_{j=1}^{n} \prod_{k=1}^{m} \left( 4\cos^2 \frac{\pi j}{2n + 1} + 4\cos^2 \frac{\pi k}{2m + 1} \right). \qquad (11.2)$$

This strange formula hides some very nice properties of these numbers. For example, in the special case of square boards ($m = n$), the number of domino tilings is always the $n$th power of two times an odd square. The first five of these numbers, $N(0, 0)$ up to $N(4, 4)$, can thus be written:

$$1, 2 = 2^1, 36 = 2^2 3^2, 6728 = 2^3 29^2, 12988816 = 2^4 901^2.$$

---

**EXERCISES**

159. Prove that $N(2, 2) = 36$ by using the formula of Kasteleyn, Fisher, and Temperley. (Difficulty rating: 1)

160. The old puzzle: Why can't the standard eight-by-eight chessboard be tiled by dominoes if two diagonally opposite corners are removed? (Difficulty rating: 2)

161. Show that however you choose one black and one white square to remove from a standard chessboard, what remains can always be tiled by dominoes. (Difficulty rating: 2)

162. Show that no staircase-shaped Ferrers board can be domino tiled. (Difficulty rating: 2)

163. Find both a necessary and sufficient condition for when a given Ferrers board can be tiled by dominoes. (Difficulty rating: 3)

---

## 11.6 The arctic circle theorem

An *Aztec diamond* is the region obtained from four staircase shapes of the same height by gluing them together along the straight edges. The Aztec diamonds of size 1, 2, and 3 are:

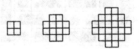

Obviously there are two domino tilings of the Aztec diamond of size 1: ☐ and ☐. Similarly, for the Aztec diamond of size 2, it is easy to construct all domino tilings and verify that there are eight of them. In general, there are $2^{m(m+1)/2}$

domino tilings of the Aztec diamond of size $m$, a formula that was found by four then-young mathematicians, Noam Elkies, Greg Kuperberg, Michael Larsen, and Jim Propp, in 1992. Thus, the formula is simpler than for the seemingly more natural square chessboards (where a certain power of two was multiplied by an odd square).

Given a domino tiling of an Aztec diamond, we shall assign an arrow for every tile. Mark with a dot the midpoint of the northernmost edge of the diamond. Then put dots at all points in the Aztec diamond reachable from the first dot via an even number of steps downward and sideways. Now every domino will have a dot at the midpoint of exactly one of its two long sides. Put an arrow in every domino, pointing toward the long-side dot.

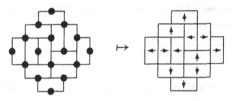

Note that the arrows of two dominoes sharing a side of length two must necessarily point in opposite directions. If the arrows point toward each other, then the two dominoes constitute a *bad block*, otherwise, a *good block*. The two kinds of blocks play an important role in a stochastic procedure to generate a random domino tiling of the Aztec diamond. The process is called *iterated shuffling*, wherein each step of a domino tiling of an Aztec diamond is extended to a tiling of the next larger Aztec diamond. Start from the empty diamond.

**Shuffling.** Each step consists of three phases. First, remove all bad blocks.

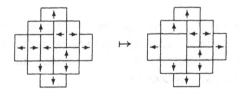

Second, slide every domino one step in the direction of its arrow.

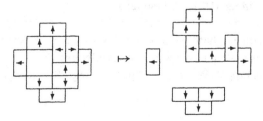

Third, create good blocks (randomly oriented either horizontally or vertically) in all two-by-two holes in order to tile the next biggest Aztec diamond. (Bigger vacancies can always be divided up in a unique way into two-by-two holes.) In the example, there are five holes to fill.

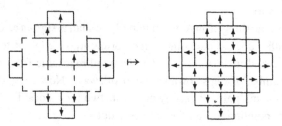

It can be shown that shuffling always works and that any tiling can be obtained by shuffling (see exercises below).

---

## EXERCISES

164. Show that no tiles ever overlap after the sliding phase.(Difficulty rating: 2)

165. How do we know that the arrows are consistent with a dot placement after the sliding phase? (Difficulty rating: 2)

166. Why is it true that "bigger vacancies can always be divided up in a unique way into two-by-two holes"? (Difficulty rating: 2)

167. Shuffling can be reversed, that is, from any domino tiling of an Aztec diamond, we can remove the good blocks and slide all tiles in the opposite direction of the arrows, and finally insert bad blocks to obtain a tiling of the next smallest Aztec diamond. Try to outline a proof of this statement, that is, break it down into smaller pieces that must be verified. (Difficulty rating: 3)

168. Given the above information about the shuffling procedure, deduce that in $m$ steps, it generates random domino tilings with uniform probability $2^{-m(m+1)/2}$. (Hence, there are $2^{m(m+1)/2}$ domino tilings of the Aztec diamond of size $m$.) (Difficulty rating: 3)

169. Draw the poset of all tilings of Aztec diamonds of size 0, 1, and 2, where one tiling precedes another if the latter can be reached from the former by iterated shuffling. (Difficulty rating: 1)

---

When the iterated shuffling procedure is run, a very particular pattern emerges. Here is a sample random domino tiling of the Aztec diamond of size 50 (from Jim Propp's Web page): http://arxiv.org/PS_cache/math/pdf/9801/9801068.pdf

We see a circle of chaotically oriented tiles flanked by four regions of completely well-ordered tiles. Jim Propp coined the terms *Arctic* (or *frozen*) zones for the well-ordered regions, *temperate* zone for the inner chaos, and *arctic circle* for the boundary between the zones. More precisely, the northern arctic zone consists of the northern connected component of tiles with arrows directed to the north, and similarly for the other three arctic zones. Thus the following Aztec diamond domino tiling

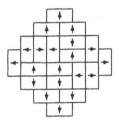

has the arctic regions

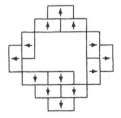

We can now state the arctic circle theorem.

**Theorem 16 (The arctic circle theorem)** *As m tends to infinity, the fraction of domino tilings of the Aztec diamond of size m whose arctic circle deviates significantly from a circle shape tends to zero.*

For more details about the amazing arctic circle theorem, the reader is referred to the original paper of Jockusch, Propp, and Shor from 1995. Here we present the basic ideas that relate the arctic circle theorem to our previous discussion about (type II) randomly growing Ferrers boards – or rather, we will use Ferrers graphs. Take the dots of the arrows in the northern arctic zone, tilt the diagram by 45 degrees, and a Ferrers graph appears!

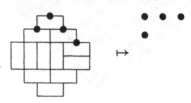

Let us see what happens to the northern arctic zone while shuffling. First, no arctic domino belongs to a bad block, so they survive the destruction phase. Second, all dominoes in the northern arctic zone slide north in the sliding phase (while the boundary dominoes do not slide north), preserving the shape of the Ferrers graph. Third, the only way the northern arctic zone can grow is by creation of horizontally oriented good blocks at the boundary. Holes at the boundary will appear precisely at the inner corners of the Ferrers graph (below marked by empty circles), and for each inner corner, there is a chance of 1/2 that the new good block will be oriented horizontally so that a new domino is included in the northern region, thus filling the inner corner with a new dot.

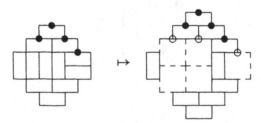

A consequence of the above argument is that the boundaries of the four arctic regions will be shaped as the contours of suitably tilted (type II) randomly growing Ferrers graphs, which according to Jockusch, Propp, and Shor, will tend to quarter-circles. Together, the four quarter-circles form the arctic circle!

---

**EXERCISE**

170. Explore different rules to govern the process of hollowing a square board by starting with a small hole in the middle and then repeatedly removing random squares at the corners of the boundary of the hole. Does any set of rules seem to lead to a circular hole? A square hole? (Difficulty rating: 3)

# Chapter 12
## Musings

In this last chapter, we want to inspire you to further studies of integer partitions!

---

**Highlights of this chapter**
- What have we left out?
- Where can you go to undertake new explorations?
- Where can you study the history of partitions?
- Are there any unsolved problems left?

---

## 12.1 What have we left out?

So here we are at the end of our introduction to the theory of partitions. Along with showing you some of the more fascinating aspects of this venerable subject, we have often implied that this is just the beginning. Many readers may wonder what's next.

### The exact formula for $p(n)$

Well, we did not show you the full, exact formula for $p(n)$, which, in fact, is

$$p(n) = \frac{1}{\pi\sqrt{2}} \sum_{k=1}^{\infty} A_k(n) k^{\frac{1}{2}} \left[ \frac{d}{dx} \frac{\sinh\left(\frac{\pi}{k}\left(\frac{2}{3}\left(x - \frac{1}{24}\right)\right)^{\frac{1}{2}}\right)}{\left(x - \frac{1}{24}\right)^{\frac{1}{2}}} \right]_{x=n},$$

where $A_k(n)$ is an explicitly given finite sum of $24k$th roots of unity. Just what you expected it to be! (Only kidding.)

121

In order to prove such a formula, one must employ quite subtle techniques from the theory of functions of a complex variable. All work of this nature relies heavily on the theory of modular forms and functions. For further reading, we suggest *The Theory of Partitions* by G. E. Andrews or *Modular Functions in Analytic Number Theory* by M. I. Knopp.

## $q$-Hypergeometric series

You have also seen a variety of formulas like

$$\sum_{n=0}^{\infty} \frac{q^{n^2}}{(1-q)^2(1-q^2)^2\cdots(1-q^n)^2} = \prod_{n=1}^{\infty} \frac{1}{1-q^n},$$

or

$$\sum_{n=0}^{\infty} \frac{q^{n^2}}{(1-q)(1-q^2)\cdots(1-q^n)} = \prod_{n=0}^{\infty} \frac{1}{(1-q^{5n+1})(1-q^{5n+4})}.$$

These are special cases of much more general theorems from the study of $q$-hypergeometric series. For example, the first formula above is a special case of

$$\sum_{n=0}^{\infty} \frac{(a-1)(a-q)\cdots(a-q^{n-1})(b-1)(b-q)\cdots(b-q^{n-1})c^n}{(1-q)(1-q^2)\cdots(1-q^n)(1-c)(1-cq)\cdots(1-cq^{n-1})}$$
$$= \prod_{n=0}^{\infty} \frac{(1-caq^n)(1-cbq^n)}{(1-cq^n)(1-abcq^n)}.$$

For further reading on this topic, the most complete source is *Basic Hypergeometric Series* by G. Gasper and M. Rahman. Some of the partition theoretic aspects of $q$-hypergeometric series may be found in the booklet *q-Series* by G. E. Andrews.

## Partitions entering other subjects

The theory of partitions also arises in combinatorial and algebraic problems in mysterious ways. For example, how many $k$-dimensional subspaces are there in an $n$-dimensional vector space defined over a finite field of $q$ elements, where $q$ is a power of a prime? The answer is

$$\begin{bmatrix} n \\ k \end{bmatrix},$$

the Gaussian polynomial that we met in Chapter 6. This and numerous other ties with combinatorial and algebraic ideas are discussed in Chapters 9 and 13 of the *Theory of Partitions* (listed above). For more advanced work, the reader is referred to *Symmetric Functions and Hall Polynomials* by I. G. Macdonald, *The Symmetric Group* by B. Sagan, and the prize-winning modern classic *Enumerative Combinatorics* by R. P. Stanley.

One of the more surprising aspects of the theory of partitions is its applicability to seemingly unrelated subjects. The book *q-Series* cited above has as its full title *q-Series: Their Development and Application in Analysis, Number Theory, Combinatorics, Physics and Computer Algebra*. As is clear from this baroque title, there are many points of contact with a variety of topics.

Henry Mann won the Cole Prize from the American Mathematical Society for his proof of the Artin conjecture. However, if you consult the Science Citation Index, you will find his paper *On a Test of Whether One of Two Random Variables Is Stochastically Larger Than the Other* with D. R. Whitney is cited many more times than his prize winner. This paper's sole object is to use combinatorial constructs equivalent to partitions enumerated by the Gaussian polynomial to provide the moments for the Wilcoxon distribution in nonparametric statistics.

## 12.2 Where can you go to undertake new explorations?

Obviously, the citations in the previous section all lead to studies beyond the introduction provided by this book. However, there is one avenue to the study of partitions that we have not mentioned: computers. One of the really exciting developments recently has been the use of computer algebra in partitions research.

### Omega

Among the more user-friendly packages is Omega, developed by G. E. Andrews, P. Paule, and A. Riese (2001) and available from

http://www.risc.uni-linz.ac.at/

This program is based on the fact that there is an algorithm for obtaining generating functions for partitions with rather complicated inequalities on the summands. For example, let $\Delta(n)$ denote the number of noncongruent triangles with integer sides and perimeter $n$. Thus,

$$\sum_{n=0}^{\infty} \Delta(n)q^n = \sum_{\substack{a_1 \geq a_2 \geq a_3 \geq 0 \\ a_2+a_3 > a_1}} q^{a_1+a_2+a_3}.$$

If these conditions on the summands are given to Omega as input, the immediate response from Omega is

$$\sum_{n=0}^{\infty} \Delta(n)q^n = \frac{q^3}{(1-q^2)(1-q^3)(1-q^4)}.$$

Of course, one can derive this formula without Omega; however, Omega produces the result automatically.

Another good example is Problem B3 on the 2000 Putnam Examination which essentially boiled down to finding a closed form for

$$S(x, y) = \sum_{\substack{m,n \geq 0 \\ 2m \geq n \\ 2n \geq m}} x^m y^n, \qquad |x| < 1, |y| < 1.$$

Omega tells us instantaneously that

$$S(x, y) = \frac{1 + xy + x^2y^2}{(1 - x^2y)(1 - y^2x)}.$$

## 12.3 Where can one study the history of partitions?

As we have indicated, Euler really kicked off the study of partitions in the eighteenth century. For a complete account, albeit without ingratiating editorial comment, of the history of partitions up to 1920, the place to go is Chapter 3 of *History of the Theory of Numbers*, Vol. II, by L. E. Dickson.

There will be a more leisurely but less complete account in the forthcoming book, *History of Combinatorics*, by Robin Wilson.

Also, there have been useful surveys of various parts of the subject in the past few decades, in particular, *Partition Identities – from Euler to the Present* by H. L. Alder and *Partition Identities* by G. E. Andrews.

Finally the history of partitions would be incomplete without noting the amazing work of the Indian genius, Srinivasa Ramanujan (1927) and its related five volume explication by Bruce Berndt (1985, 1989, 1991, 1994, 1998).

## 12.4 Are there any unsolved problems left?

There are many such problems. We have already presented Alder's conjecture in Section 4.3. We shall cite three that require no more knowledge than that contained in the previous chapters.

### Borwein's problem

Let $B_e(N; n)$ (resp. $B_o(N; n)$) denote the number of partitions of $n$ into an even (resp. odd) number of distinct nonmultiples of 3 each $< 3N$. Prove that for all positive integers $N$ and $n$,

$$B_e(N; n) - B_o(N; n)$$

is nonnegative if $n$ is a multiple of 3 and nonpositive otherwise.

Background on this conjecture may be found in *On a Conjecture of Peter Borwein* by G. E. Andrews (1995).

### Leibniz's problem

Leibniz (assuming 1 to be a prime) suggested that $p(n)$ might always be prime in light of the fact that $p(n)$ is prime for $n \leq 6$. Of course, $p(7) = 15$, so we must modify this suggestion into the following problem: Prove that $p(n)$ is prime for infinitely many $n$. Ono (2000) has proved that every prime divides at least one value of $p(n)$.

### Parity of $p(n)$

In recent years, Ken Ono has made dramatic discoveries about the divisibility properties of $p(n)$, cf. Ono (2000). Many questions are still open. For example, prove

$$\lim_{n \to \infty} \frac{\left| \{ m | m \leq n \text{ and } p(m) \text{ even} \} \right|}{n} = \frac{1}{2}.$$

In other words, prove that approximately half the values of $p(n)$ are even.

Now that you have finished this book we encourage you to take a shot at an open problem. Reading can be fun, but research can be more fun.

# Appendix A

## On the convergence of infinite series and products

All of the infinite series considered in this book can be proved to converge by two standard tests from calculus:

### The ratio test

The infinite series $\sum_{n=0}^{\infty} a_n$ converges absolutely if $\lim_{n\to\infty} \left|\frac{a_{n+1}}{a_n}\right| < 1$.

### The root test

The infinite series $\sum_{n=0}^{\infty} a_n$ converges absolutely if $\lim_{n\to\infty} |a_n|^{\frac{1}{n}} < 1$.
For example,

$$\sum_{n=0}^{\infty} \frac{z^n q^{\frac{n(n-1)}{2}}}{(1-q)(1-q^2)\cdots(1-q^n)}$$

converges absolutely, provided $|q| < 1$. To see this, we apply the ratio test:

$$\lim_{n\to\infty} \left| \frac{z^{n+1} q^{\frac{n(n+1)}{2}}}{(1-q)(1-q^2)\cdots(1-q^{n+1})} \middle/ \frac{z^n q^{n(n-1)/2}}{(1-q)(1-q^2)\cdots(1-q^n)} \right|$$

$$= \lim_{n\to\infty} \left| \frac{z\, q^n}{1-q^{n+1}} \right| = 0 \text{ provided } |q| < 1.$$

As an example of the root test, we recall that in Chapter 6, we proved that

$$\lim_{n\to\infty} p(n)^{\frac{1}{n}} = 1.$$

Hence, $\sum_{n=0}^{\infty} p(n)q^n$ converges absolutely for $|q| < 1$ because

$$\lim_{n\to\infty} |p(n)q^n|^{\frac{1}{n}} = \lim_{n\to\infty} p(n)^{\frac{1}{n}}|q| = |q|,$$

and so $|q| < 1$ will guarantee absolute convergence of the series in question.

Less familiar, owing to neglect in standard calculus courses, are infinite products.

We define

$$\prod_{n=0}^{\infty}(1+a_n) = \lim_{N\to\infty}\prod_{n=0}^{N}(1+a_n) = \lim_{N\to\infty}(1+a_0)(1+a_1)\cdots(1+a_N),$$

provided the limit exists and is *not zero*, and we say the infinite product converges.
We say $\prod_{n=0}^{\infty}(1+a_n)$ is absolutely convergent, providing $\prod_{n=0}^{\infty}(1+|a_n|)$ converges.
We shall state and prove three facts about the convergence of infinite products.

## Fact 1

If $a_n \geq 0$ for each $n$, then $\prod_{n=0}^{\infty}(1+a_n)$ and $\sum_{n=0}^{\infty}a_n$ are both convergent or both divergent.

This assertion follows immediately from the inequalities:

$$1+a_1+a_2+\cdots+a_N \leq \prod_{n=1}^{N}(1+a_n) \leq e^{a_0+a_1+\cdots+a_N}.$$

The left-hand inequality follows by mathematical induction on $N$, and the right-hand inequality is a direct consequence of the fact that for all real $x$,

$$1+x \leq e^x.$$

## Fact 2

If $1 > a_n \geq 0$ for each $n$, then $\prod_{n=0}^{\infty}(1-a_n)$ and $\sum_{n=0}^{\infty}a_n$ are both convergent or both divergent.

The proof of this assertion is slightly subtler than the previous one. This is because we must take into account that portion of the definition of an infinite product requiring that the limiting value not be zero. The idea, however, is much the same. Now we use the analogous inequalities: for $m \geq N$,

$$1-a_N-a_{N+1}\cdots-a_m \leq (1-a_N)(1-a_{N+1})\cdots(1-a_m) \leq e^{-a_N-a_{N+1}-\cdots-a_m}.$$

So on the one hand, if $\sum_{n=0}^{\infty}a_n$ converges, then we can find $N$ so that $\sum_{n=N}^{\infty}a_n < \frac{1}{2}$. This means that the non-increasing sequence of partial products $\prod_{n=0}^{m}(1-a_n)$ is (for $m \geq n$) bounded below by

$$\frac{1}{2}\prod_{n=0}^{N-1}(1-a_n),$$

and so converges to a positive limit, that is, the infinite product also converges.

On the other hand, if the infinite product converges, then there exists a positive number $c$ so that

$$0 < c < (1-a_0)(1-a_1)\cdots(1-a_N) \leq e^{-a_0-a_1\cdots-a_N}$$
$$\therefore \quad \log c \leq -a_0 - a_1 \cdots - a_N$$

or $a_0 + a_1 + \cdots + a_N \leq \log \frac{1}{c}$. Thus $\sum_{n=0}^{\infty} a_n$ converges because the partial sums form a bounded increasing sequence.

## Fact 3

If $|a_n| < 1$ for each $n$ and if $\prod_{n=0}^{\infty}(1 + |a_n|)$ converges, then $\prod_{n=0}^{\infty}(1 + a_n)$ converges.
    To see that this third fact is true, we define

$$P_N = \prod_{n=0}^{N}(1 + |a_n|)$$

and

$$p_N = \prod_{n=0}^{N}(1 + a_n).$$

First of all,

$$
\begin{aligned}
|p_N - p_{N-1}| &= |(1 + a_1)(1 + a_2)\cdots(1 + a_{N-1})a_N| \\
&\leq (1 + |a_1|)(1 + |a_2|)\cdots(1 + |a_{N-1}|)|a_N| \\
&= P_N - P_{N-1}.
\end{aligned}
\tag{A.1}
$$

It is now an easy exercise in mathematical induction to prove that for $R > S$,

$$|p_R - p_S| \leq P_R - P_S.$$

Hence, convergence of the sequence $P_N$ forces convergence of the sequence $p_n$. All that remains is a proof that $\lim_{n \to \infty} p_n \neq 0$. But this follows from

$$|p_N| \geq (1 - |a_0|)(1 - |a_1|)\cdots(1 - |a_N|)$$

and the facts that

(i) $\sum_{n=0}^{\infty}|a_n|$ converges,
(ii) $|a_n| < 1$, and
(iii) by Fact 2, $\prod_{n=0}^{\infty}(1 - |a_n|)$ converges to a positive limit, which means all partial products thereof are bounded below by a positive constant.

# Appendix B

## References

H. L. Alder, Partition identities–from Euler to the present, *Amer. Math. Monthly* **76** (1969) 733–746.

K. Alladi, The method of weighted words and applications to partitions, Number Theory, S. David ed., Cambridge University Press, Cambridge 1995.

K. Alladi, G. E. Andrews, and A. Berkovich, A new four parameter $q$-series identity and its partition implications, Invent. Math. **153** (2003), 231–260.

G. E. Andrews, On radix representation and the Euclidean algorithm, *Amer. Math. Monthly* **76** (1969a) 66–68.

G. E. Andrews, Two theorems of Euler and a general partition theorem, *Proc. Amer. Math. Soc.* **20** (1969b) 499–502.

G. E. Andrews, On a partition problem of H. L. Alder, *Pac. J. Math.* **36** (1971a) 279–284.

G. E. Andrews, The use of computers in search of identities of the Rogers-Ramanujan type, *Computers in Number Theory*, A. O. L. Atkin and B. J. Birch, eds., Academic Press, New York, (1971b) 377–387.

G. E. Andrews, A combinatorial proof of a partition function limit, *Amer. Math. Monthly* **76** (1971c) 276–278.

G. E. Andrews, Partition identities, *Advances in Math.* **9** (1972) 10–51.

G. E. Andrews, Partition ideals of order 1, the Rogers-Ramanujan identities and computers, *Proc. Sminaire Dubreil (algbre) 19ieme anne* **20** (1975) 1–16.

G. E. Andrews, Partitions and Durfee dissection, *Amer. J. Math.* **101** (1979) 735–742.

G. E. Andrews, On a partition theorem of N. J. Fine, J. Natl. Acad. Math. India **1** (1983) 105–107.

G. E. Andrews, Generalized Frobenius partitions, *Memoirs AMS* **49** (1984) iv + 44.

G. E. Andrews, *q-Series: Their Development and Application in Analysis, Number Theory, Combinatorics, Physics and Computer Algebra*, CBMS Regional Conference Series, No. 66, Amer. Math. Soc., Providence, R.I., 1986.

G. E. Andrews, On a conjecture of Peter Borwein, *J. Symbolic Computation* **20** (1995) 487–501.

G. E. Andrews, *The Theory of Partitions*, Cambridge Mathematical Library, Cambridge University Press, Cambridge, U.K., 1998.

G. E. Andrews, P. Paule, and A. Riese, MacMahon's partition analysis: The Omega package, *Europ. J. Combinatorics* **22** (2001) 887–904.

G. E. Andrews, Partitions: At the interface of $q$-series and modular forms, *Ramanujan J.* **7** (2003), 385–400.

A. O. L. Atkin and H. P. F. Swinnerton-Dyer, Some properties of partitions, *Proc. London Math. Soc.* **4** (1953) 84–106.

A. Berkovich and B.M. McCoy, Rogers-Ramanujan identities: A century of progress from mathematics to physics, Doc. Math. J. DMV, Extra Volume ICM 1998, III, 163–172.

B. Berndt, Ramanujan's Notebooks, Parts I–V, Springer, Berlin, 1985, 1989, 1991, 1994, 1998.

M. Bousquet-Mélou and K. Eriksson, Lecture hall partitions, *Ramanujan J.* **1** (1997a) 101–110.

M. Bousquet-Mélou and K. Eriksson, Lecture hall partitions 2, *Ramanujan J.* **1** (1997b) 165–185.

M. Bousquet-Mélou and K. Eriksson, A refinement of the lecture hall partition theorem, *J. Comb. Th. (A)* **86** (1999) 63–84.

D. M. Bressoud, A new family of partition identities, Pacific J. Math. **77** (1978) 71–74.

D. Bressoud, Some identities for terminating $q$-series, *Math. Proc. Cambridge Phil. Soc.* **89** (1981) 211–223.

L. Carlitz, Rectangular arrays and plane partitions, *Acta Arith.* **13** (1967) 29–47.

R. Chapman, A new proof of some identities of Bressoud, *Int. J. Math. and Math. Sciences* **32** (2002) 627–633.

L. E. Dickson, *History of the Theory of Numbers, Vol. 2, Diophantine Analysis*, Chelsea, New York, 1952.

F. J. Dyson, Some guesses in the theory of partitions, *Eureka (Cambridge)* **8** (1944) 10–15.

N. Elkies, G. Kuperberg, M. Larsen, and J. Propp, Alternating-sign matrices and domino tilings, J. Alg. Combinatorics **1** (1992) 111–132.

G. Gasper and M. Rahman, *Basic Hypergeometric Series*, Cambridge University Press, Cambridge, U.K., 1990.

G. H. Hardy, *Ramanujan*, Cambridge University Press, Cambridge, U.K., 1940 (Reprinted: Chelsea, New York, 1959).

W. Jockusch, J. Propp and P. Shor, Random domino tilings and the arctic circle theorem, available from http://www.math.wisc.edu/~propp/articles.html

R. Kanigel, *The Man Who Knew Infinity*, Washington Square Press, New York, 1991.

D. Kim and A. J. Yee, A note on partitions into distinct and odd parts, *Ramanujan J.* **3** (1999) 227–231.

M. I. Knopp, *Modular Functions in Analytic Number Theory*, Chelsea, New York, 1993.

I. G. Macdonald, *Symmetric Functions and Hall Polynomials*, Oxford University Press, Oxford, U.K., 1979.

P. A. MacMahon, Memoir on the theory of partition of numbers I, *Phil. Trans.* **187** (1897) 619–673 (Reprinted: *Collected Papers*, 1026–1080).

P. A. MacMahon, *Combinatory Analysis, Vol. 2*, Cambridge University Press, Cambridge, U.K., 1916 (Reprinted: Chelsea, New York, 1960).

H. B. Mann and D. R. Whitney, On a test of whether one of two random variables is stochastically larger than the other, *Annals of Math. Statistics* **18** (1947) 50–56.

K. Ono, Distribution of the partition function modulo $m$, *Annals of Math.* **151** (2000) 293–307.

B. Pittel, On a likely shape of the random Ferrers diagram, *Adv. Appl. Math.* **18** (1997) 432–488.

S. Ramanujan, *Collected Papers*, Cambridge University Press, Cambridge, U.K., 1927 (Reprinted: Chelsea, New York, 1962).

H. Rost, Non-equilibrium behavior of a many particle system: density profile and local equilibria, Probab. Theory Relat. Fields **58** (1981) 41–53.

B. Sagan, *The Symmetric Group*, Wadsworth, Pacific Grove, Calif., 1991.

I. Schur, Ein Beitrag zur additiven Zahlentheorie und zue Theorie der Kettenbrüche, *S.-B. Preuss. Akad. Wies. Phys.-Math. Kl.*, pp. 302–321 (Reprinted: *Ges. Abhandlungen, Vol. 2*, pp. 117–136).

I. Schur, Zur additiven Zahlentheorie, *S.-B. Preuss. Akad. Wies. Phys.-Math. Kl.*, pp. 488–495 (Reprinted: *Ges. Abhandlungen, Vol. 3*, pp. 43–50).

R. P. Stanley, *Enumerative Combinatorics, Vol. 1*, Wadsworth, Pacific Grove, Calif., 1986.

R. P. Stanley, *Enumerative Combinatorics, Vol. 2*, Cambridge, U.K., 1999.

M. V. Subbarao, Partition theorems for Euler pairs, *Proc. Amer. Math. Soc.* **28** (1971a) 330–336.

M. V. Subbarao, On a partition theorem of MacMahon–Andrews, Proc. Amer. Math. Soc. **27** (1971b) 449–450.

J. J. Sylvester, A constructive theory of partitions arranged in three acts, an interact and an exodion, *Amer. J. Math.* **5** (1884) 251–330, **6** (1886) 334–336 (Reprinted: *Collected Papers, Vol. 4*, 1–83).

A. Yee, On the combinatorics of lecture hall partitions, Ramanujan J. **5** (2001) 247–262.

A. Yee, On the refined lecture hall theorem, Discr. Math. **248** (2002) 293–298.

# Appendix C

## Solutions and hints to selected exercises

**3** An obvious bijection proving the equality $p(n \mid$ even parts$) = p(n/2)$: For any partition of $n$ into even parts, replace every part with a part of half the size. An obvious bijection proving the equality $p(n/2) = p(n \mid$ even number of each part$)$: For any partition of $n/2$, replace every part by two parts of the same size.

**4** Every step in the splitting/merging procedure changes the number of odd parts by an even number ($+2$ if an even part is split into two odd parts, $-2$ if two odd parts are merged, and $0$ otherwise). Hence, the parity (odd or even) of the number of odd parts is the same through the entire procedure.

**7** Let $M$ be the set of all positive integers that are either a power of two or three times a power of two. Then Theorem 1 says that $p(n \mid$ distinct parts in $M)$ equals $p(n \mid$ parts in $\{1, 3\})$. Obviously there are $\lfloor n/3 \rfloor + 1$ ways of choosing the number of 3:s in such a partition, and then there is a unique way of completing the partition with 1:s.

**8** If $n$ is the smallest integer that lies in one set, say $M$, but not in the other, say $M'$, then $p(n \mid$ distinct parts in $M) = 1 + p(n \mid$ distinct parts in $M')$, for the partitions counted are identical except for the partition consisting of the part $n$ only.

**9** Let $N'$ be the set consisting of those elements in $N$ that are not a power of two times some other element in $N$. Let $M'$ be the set containing all elements of $N'$ together with all their multiples of powers of two. Then, according to Theorem 1, the pair $(N', M')$ is an Euler pair. The element $2^k a$ of $N$ is the smallest element not in $N'$, so when trying to construct a set $M$ such that $(N, M)$ is an Euler pair, we are forced to follow exactly the construction of $M'$ up to $2^k a$. For this element, we fail, because $2^k a$ is included already in $M'$, so there is no possibility to obtain more partitions of $2^k a$ with distinct parts in $M$ than the corresponding partitions with distinct parts in $M'$, whereas there is (exactly) one more partition of $2^k a$ with parts in $N$ than with parts in $N'$, namely, the partition consisting of $2^k a$ only.

**10** The condition $2M \subset M$ says that for each element in $M$, every power of two times that element is also in $M$. The condition $N = M - 2M$ says that $N$ consists of all elements in $M$ that are not a power of two times any other element in $M$. Hence, $N$ is a set of integers such that no element of $N$ is a power of two times an element of $N$, and $M$ is the set containing all elements of $N$ together with all their multiples of powers of two, so $(N, M)$ is an Euler pair.

Conversely, if $(N, M)$ is an Euler pair, then $2M \subset M$ and $N = M - 2M$.

**12** A Ferrers graph is a collection of rows of equidistant dots such that the left margin is straight and every row (except the last one) is at least as long as the row below it.

**13** Hint: Two adjacent outer corners determine the position of the inner corner in between.

**14** Hint: Two adjacent inner corners determine the position of the outer corner in between.

**15** Hint: Every inner corner will, after enlargement of the partition, yield a new inner corner in the next column. In addition, we always have an inner corner at the bottom of the first column.

**16** (a) $6 + 4 + 2$, (b) $2 + 1 + 1 + 1 + 1 + 1 + 1$, (c) $5 + 4 + 2 + 2 + 1$

**18** A partition has $\leq m$ parts precisely if the top row of its Ferrers graph has length $\leq m$. A partition has all parts $\leq m$ precisely if the first column of its Ferrers graph has length $\leq m$. Conjugation is an obvious bijection between these Ferrers graphs.

**24** Partitions that are not self-conjugate come in conjugate pairs and therefore do not affect the parity of $p(n)$. Hence, $p(n)$ is odd if and only if $p(n \mid \text{self-conjugate})$ is odd; and by Eq. (3.4), we have $p(n \mid \text{self-conjugate}) = p(n \mid \text{distinct odd parts})$.

**25** Every partition of $n$ with Durfee side $= j$ can be uniquely decomposed into the Durfee square (of size $j^2$), a Ferrers board below (of, say, size $m$) with rows of length at most $j$, and a Ferrers board to the right (of size equal to the remaining number of dots, that is $n - j^2 - m$) with columns of length at most $j$.

**27** The Fibonacci sequence starts $0, 1, 1, 2, 3, 5, 8, 13, 21, 34$.

**29** The identity $F_n = F_{n-1} + F_{n-3} + F_{n-5} + \ldots$ is true, by inspection, for $n = 2$ and $n = 4$. For even $n > 4$, it follows by induction, for then $F_n = F_{n-1} + F_{n-2} = F_{n-1} + (F_{n-3} + F_{n-5} + \ldots)$.

**30** Hint: Compositions of $n$ into 1s and 2s come in two categories: those where the last term is a 1, and those where the last term is a 2.

**31** The first time the value of the partition function differs from the Fibonacci number is for $n = 5$: $p(5) = 7 \neq 8 = F_5$. This is because 5 is the smallest value of $n$ such that there exists a partition of $n - 2$ the smallest non-1-part of which is less than $2 + \#1$-parts; this partition being $2 + 1$.

**32** Hint: $\tau^n = \tau^{n-1} + \tau^{n-2}$.

**33** There are $C_4 = \binom{8}{4}/5 = 14$ partitions fitting inside a staircase of height 4: $3 + 2 + 1, 3 + 2, 3 + 1 + 1, 3 + 1, 3, 2 + 2 + 1, 2 + 2, 2 + 1 + 1, 2 + 1, 2, 1 + 1, 1 + 1, 1$, and the empty partition.

**35** The odd parts will end up as rows at the bottom of the graph. Say that there are $k$ odd parts. Then the smallest even part will, by the construction of the graph, have $2k + 1$ dots to the left of the line (and at least one dot to the right of the line). Hence, each even part is greater than twice the number of odd parts.

**36** Take any partition into distinct parts with each even part greater than twice the number of odd parts. Arrange the rows such that the even rows come first, in decreasing order, followed by the odd rows in decreasing order. Adjust the left margin to a slope of two dots extra indentation per row. Draw a vertical line in such a way that the last row has one dot to the left of this line. We must show that all rows reach the vertical line. Distinct odd parts differ by at least 2, so they must reach the line. With $k$ odd parts, the smallest even part is at least $2k + 2$, so it reaches the line too, and then so do the larger even parts. If we rearrange the rows to the right of the line in descending order, we obtain a partition into super-distinct parts. Obviously, this procedure inverts the previous procedure.

**39** Hint: Move the smallest part within the pair, a lá Franklin!

**41** The Franklin transformation always changes the parity of the largest part.

**43** Sylvester's bijection takes any partition of $n$ that is *not* into odd distinct parts and pairs it up with another such partition, where the number of parts differs by one. Hence, it tells us that among all partitions of $n$ that are *not* into odd distinct parts, exactly half have an even number of parts. Therefore, to compute the difference

$$p(n \mid \text{even number of parts})$$
$$-p(n \mid \text{odd number of parts}),$$

we need only consider partitions into odd distinct parts. Since the sum of an odd number of odd parts is always odd, the parity of the number of parts of these partitions equals the parity of $n$. In other words, all the partitions into odd distinct parts will contribute to the positive term if $n$ is even, and to the negative term if $n$ is odd.

**44** Sylvester's bijection also changes the number of even parts by exactly one, so to compute the difference

$$p(n \text{ number of even parts is even})$$
$$-p(n \text{ number of even parts is odd})$$

we need again only consider partitions into odd distinct parts. But the number of even parts in such partitions is always even, namely, zero.

**45** An example of a partition identity that is not of type (4.1) is $p(n \mid \text{self-conjugate}) = p(n \mid \text{distinct odd parts})$.

**46** Since $\lambda_2 \geq 2\lambda_1$, we can construct a partition of $n$ into $\lambda_1$ 3s and $(\lambda_2 - 2\lambda_1)$ 1s. Clearly this contruction is invertible, and hence a bijection.

**48** From a partition of $n$ as $\lambda_2 + \lambda_1$, where $\frac{3}{2}\lambda_1 \geq \lambda_2 \geq \lambda_1 \geq 0$, we can construct a partition of $n$ into $(\lambda_2 - \lambda_1)$ 5s and $(3\lambda_1 - 2\lambda_2)$ 2s. Conversely, a partition into $e_5$ 5s and $e_2$ 2s is mapped back to a partition $\lambda_2 + \lambda_1$, where $\lambda_2 = 3e_5 + e_2$ and $\lambda_1 = 2e_5 + e_2$, which clearly satisfies the inequalities.

**50** $N = \{2, 3, 7, 8, \ldots\}$, that is, all positive integers congruent to 2 or 3 modulo 5.

**51** $N = \{1, 4, 7, 9, 12, 15, \ldots\}$, that is, all positive integers congruent to 1, 4, or 7 modulo 8.

**52** $N = \{3, 4, 5, 11, 12, 13, \ldots\}$, that is, all positive integers congruent to 3, 4, or 5 modulo 8.

**53** $N = \{1, 5, 7, 11, \ldots\}$, that is, all positive integers congruent to 1 or 5 modulo 6.

**56** For $d = 3$, we can construct the first four elements of $N$ to be $\{1, 5, 7, 9\}$, but then for $n = 10$, the construction breaks down since there are already five partitions of 10 into parts in $N$ ($1^{10}, 5^1 1^5, 5^2, 7^1 1^3, 9^1 1^1$) but only four partitions of 10 into 3-distinct parts (10, $9 + 1, 8 + 2, 7 + 3$). For $d = 4$, the construction starts $N = \{1, 6, 8, 10, 15\}$ and breaks down for $n = 16$, which has nine partitions into parts in $N$ but only eight partitions into 4-distinct parts.

**59** $N$ contains all odd numbers that are not divisible by three, and $M$ contains all numbers that are not divisible by three. The property of divisibility by three is invariant under the merging-splitting process of Chapter 2.

**63** $N$ is the set of positive integers congruent to $\pm 1$ modulo 5, by the first Rogers-Ramanujan identity!

**98**

$$\sum_{m=0}^{N} q^{m(m+1)/2} \begin{bmatrix} N \\ m \end{bmatrix}$$
$$= (1+q)(1+q^2)\cdots(1+q^N),$$

and

$$\sum_{m=0}^{N} (-1)^m q^{m(m+1)/2} \begin{bmatrix} N \\ m \end{bmatrix}$$
$$= (1-q)(1-q^2)\cdots(1-q^N).$$

**111** 12 is the only lecture hall partition of length 1. For length 2, we add $1 + 11, 2 + 10, 3 + 9$, and $4 + 8$. For length 3, we also include $1 + 2 + 9, 1 + 3 + 8, 1 + 4 + 7$, and $2 + 4 + 6$. Finally, for length 4, we add $1 + 2 + 3 + 6, 2 + 3 + 7$, and $5 + 7$.

**112** The assertion $\mathcal{L}_N \subseteq \mathcal{L}_{N+1}$ is equivalent to the assertion that the inequality $\frac{\lambda_k}{k} \leq \frac{\lambda_{k+1}}{k+1}$ implies the inequality $\frac{\lambda_k}{k+1} \leq \frac{\lambda_{k+1}}{k+2}$, which follows immediately from the simple fact that $\frac{k}{k+1} \leq \frac{k+1}{k+2}$ for all natural numbers $k$.

**113** Consecutive parts satisfy the inequality $\lambda_k \leq \frac{k}{k+1}\lambda_{k+1}$, and for nonzero parts the right-hand expression is strictly less than $\lambda_{k+1}$. Hence, nonzero parts are distinct.

**116** Hint: This is Exercise 2 of Chapter 4.

**119** There are four partitions of 19 into three odd parts smaller than 10, namely, $9 + 9 + 1, 9 + 7 + 3, 9 + 5 + 5$, and $7 + 7 + 5$. There are also four lecture hall partitions $\lambda$ of 19 of length 5 with $s(\lambda) = 3$, namely, $8 + 11, 1 + 8 + 10, 1 + 2 + 7 + 9$, and $2 + 3 + 6 + 8$.

**120** The number of partitions of $n$ into $S$ odd parts equals the number of partitions $\lambda$ of $n$ into distinct parts with $s(\lambda) = S$.

**122** $0 + 0$ and $0 + 1$.

**124** A lecture hall partition is reduced if the lecture hall property is destroyed whenever $k$ is subtracted from the $k$th part, for all $k = 1, 2, \ldots, N$. Hence, we can construct all reduced lecture hall partitions one part at a time; given the $k - 1$ first parts $\lambda_1, \ldots, \lambda_{k-1}$, we can choose the $k$th part $\lambda_k$ in $k$ ways: take the smallest possible value satisfying the lecture hall inequality $\frac{\lambda_{k-1}}{k-1} \leq \frac{\lambda_k}{k}$, then add any integer between 0 and $k - 1$.

**126** For $N = 2$, we have the blocks $1 + 2$ (of size 3) and $0 + 2$ (of size 2), whereas the reduced lecture hall partitions are $0 + 0$ (of size 0) and $0 + 1$ (of size 1). The generating function becomes

$$\frac{q^0 + q^1}{(1-q^3)(1-q^2)}$$
$$= \frac{(1-q^2)/(1-q)}{(1-q^3)(1-q^2)}$$
$$= \frac{1}{(1-q)(1-q^3)}.$$

**127** One would need to prove that the generating function for the reduced lecture hall partitions of length $N$ equals

$$\frac{(1-q^{1+2+\cdots+N})(1-q^{2+\cdots+N})\cdots(1-q^N)}{(1-q)(1-q^3)(1-q^5)\cdots(1-q^{2N-1})}.$$

We have thus reduced the problem to finding the generating function for the reduced lecture hall partitions, but this too is a nontrivial task.

**128** A bijective proof is: From a partition into odd parts $< 2N$, repeatedly merge pairs of equal parts $\leq N$. The inverse of this is to repeatedly split even parts into halves. A generating function proof would instead verify the following identity:

$$\frac{1}{(1-q)(1-q^3)\cdots(1-q^{2N-1})}$$
$$= \frac{(1+q)(1+q^2)\cdots(1+q^N)}{(1-q^{N+1})(1-q^{N+2})\cdots(1-q^{2N})}.$$

This is easily accomplished by induction over $N$ using the identity $\frac{(1+q^N)}{(1-q^{2N})/(1-q^N)} = 1$.

**139** Either the partition or its conjugate must consist of a single part.

**140** For Ferrers boards, it is obvious that if $\lambda$ is contained in $\mu$, then conjugating both boards does not affect this relation.

**141** The *lexicographic order* is the order used in dictionaries; view partitions (with parts in decreasing order) as words over the alphabet of the positive integers, and order them as they would be found in a dictionary. For example, in this order, we have $1 <_{\text{lex}} 1+1 <_{\text{lex}} 1+1+1 <_{\text{lex}} \cdots <_{\text{lex}} 2 <_{\text{lex}} 2+1 <_{\text{lex}} 2+1+1 <_{\text{lex}} \cdots <_{\text{lex}} 2+2$.

**143** The usual order $\lambda \preceq \mu$ can be described by the inequalities $\lambda_1 \leq \mu_1$, $\lambda_2 \leq \mu_2$, $\lambda_3 \leq \mu_3$, etc. Clearly the inequalities of the dominance order are implied. The converse does not hold; for example, $1+1 \trianglelefteq 2$ but $1+1 \not\preceq 2$.

**145** There are 26 standard tableaux with 5 squares.

**147** The element $n$ is either by itself (leaving $n-1$ elements to be partitioned) or paired with any one of the other $n-1$ elements (leaving $n-2$ elements to be partitioned).

**148** $f^6 = 1$, $f^{5+1} = 5$, $f^{4+2} = 9$, $f^{4+1+1} = 10$, $f^{3+3} = 5$, $f^{3+2+1} = 16$.

**150** The sum $\sum_{\lambda \vdash n} (f^\lambda)^2$ evaluates to $n!$. This is proved by the neat Robinson-Schensted-Knuth correspondence, cf. Stanley (1999).

**151** Verify that the hook length formula for a two-rowed partition $n+m$ simplifies to $(n+m)!(n+1-m)/(n+1)!m!$, and then just check the recursion by algebra.

**155** The binary sequence describes the contour of the board, with 0 denoting a step upward and 1 a step to the right. The pattern 01 is a step upward followed by a step to the right, that is, an inner corner. Adding a square to this inner corner clearly changes the pattern to 10.

**158** On a black-and-white chessboard, every domino always covers one square of each color, hence the number of white squares must equal the number of black squares in any tiling of any subset of the chessboard. Thus, removing two squares of the same color effectively destroys all possibility of a domino tiling, and diagonally opposite corners are of the same color.

**159** Hint: Find a closed path from square to neighboring square that visits all the squares of the chessboard exactly once (this is easy). Removing one black and one white square will cut this closed path into two segments, each with one black and one white end, hence of even lengths and thus tileable by dominoes.

**160** Hint: Show that staircase-shaped boards cannot have as many white as black squares.

**161** The necessary condition that there must be equally many black and white squares is in fact also sufficient for domino tileability of Ferrers boards. Sufficiency follows by induction; since the board cannot be staircase shaped (above exercise), its contour must allow removal of a domino somewhere, leaving a smaller Ferrers board satisfying the criterion.

**162** Hint: There are just a few cases of overlapping dominoes, each of which can be ruled out as impossible to appear.

**163** Regard the dots of the Aztec diamond as the visible part of an infinite grid of dots:

In the sliding phase, the top domino is moved one step, hence the whole dot pattern should be moved one step (sideways or upward, it is obviously the same for this grid). Since all dominoes also slide one step, sideways or upward, their arrows will be consistent with the dots.

**165** Proof outline:

- Prove that sliding backwards after good blocks have been removed will never create overlaps. (Easy, just check the two possible cases of overlap.)

- Verify that the new configuration must be contained in the next smallest Aztec diamond. (Easy, just check how dominoes can slide near the border.)
- Show that any vacancies in the new smallest Aztec diamond can be divided into two-by-two holes that will be filled with bad blocks.

**166** By reversibility, any domino tiling of the Aztec diamond can be reached by shuffling. An Aztec diamond of size $m$ contains $4m$ more squares (around the border) than the next smaller size, so if $b$ bad blocks are removed when shuffling from size $m - 1$ to size $m$, then $b + m$ good blocks must be added. Any of $2^b$ orientations of the bad blocks can result in any of $2^{b+m}$ orientations of good blocks, so the ratio is $2^m$, independent of $b$. Therefore, if tilings of the Aztec diamond of size $m - 1$ are generated with uniform probability $2^{-(m-1)m/2}$, then the probability of any particular domino tiling of an Aztec diamond of size $m$ is

$$2^{-(m-1)m/2} \cdot 2^{-m} = 2^{-m(m+1)/2}.$$

# Index

139